Enzymes:
Physical Principles

H. GUTFREUND

Director, Laboratory of Molecular Enzymology,
Department of Biochemistry,
University of Bristol

Wiley–Interscience

a division of John Wiley & Sons Ltd
London–New York–Sydney–Toronto

Library of Congress Catalog card No. 75-39068

ISBN 0 471 33715 3

Printed in Great Britain by
J. W. Arrowsmith Ltd.,
Winterstoke Road, Bristol BS3 2NT

Preface

It is customary to state in the Preface why one has written the book, why one has included or omitted some borderline aspects of the subject and for whom one has written the book. This is a good custom provided it introduces the potential reader to the author and his prejudices and helps him to use the book.

My main prejudice is that an understanding of physical principles is more important than the memorizing of facts. In my opinion the present examination system in biochemistry puts too much emphasis on the descriptive. Unfortunately, memorizing the problems which have been solved in the past does not provide a good basis for solving the problems of the future unless concepts and general principles are emphasized.

The book can be used at several levels. A student who has taken one-year university courses in Biochemistry and in Chemistry should be able to work through the whole text and understand most of it sufficiently well to gain a sound knowledge of those aspects of physical chemistry which are essential for the pursuit of modern biochemistry. Those readers who become encouraged by such a first reading to take the study of physical principles more seriously can carry on in a variety of ways. If they find serious gaps in their education in some aspects of physical chemistry, which they wish to remedy, they should study the textbooks or reviews referred to for further reading in that particular field. For those who wish to take up some special topic enough original papers are quoted to introduce them to their chosen topic.

A special word for those who are worried by the sight of a few simple mathematical equations. Equations are used as the simplest way to describe physical relations. It is not necessary to be able to manipulate differential equations or matrix algebra to understand or use 95 per cent. of the text. In most cases phenomena are explained in words as well as in equations. It seems likely to me that most students who have a minimum of mathematical training will get some satisfaction from working their way through all but a few of the derivations. They will suddenly see how simple it really is. Some readers will find the detail in the mathematical treatment patchy. This is done for a purpose; if the student fills in the easy steps he will understand the concepts described. If every simple step is written down in the book, this causes some inhibition of thought processes and is too much of an encouragement for memorizing even algebraic derivations.

The topics treated have been chosen to some extent with an eye for the future. It is difficult to predict what aspects of physical chemistry are going to be important in the future for the study of enzyme molecules. It is likely that a much wider view will be taken of the role of protein molecules as devices for control and transport in addition to their catalytic activity. More emphasis is placed on the quantitative treatment of specificity, recognition and protein assembly than has been the custom in books on enzyme mechanisms. A book which deals with all aspects of the physical chemistry of enzymes in the detail given here on ligand binding equilibria and modern kinetic analysis would fill about four volumes of the present size. Such detailed treatment is given to reaction mechanisms by Jencks (1969) and Bruice and Benkovic (1966), to the size and shape of macromolecules by Tanford (1962), to ionic interactions by Edsall and Wyman (1968) and to protein structure by Dickerson and Geis (1969). Clearly it would have been possible to include a brief discussion of the application of nuclear magnetic resonance to the study of enzyme reactions. I decided reluctantly that adequate treatment of this important subject would be beyond the scope of this book (see Hague, 1971, for references to this subject).

I hope that the present volume will fill a need. In the author's experience the subject-matter dealt with can be covered in a course of 24 lectures. Each lecturer involved in such a course will wish to emphasize different aspects of the subject with supplementary reading.

I have been very fortunate in my friends and collaborators during the last 25 years. I hope they will accept my thanks without being mentioned in a long list of names. A special vote of thanks goes to my colleagues in the Molecular Enzymology Laboratory at the University of Bristol, who have not only helped me to build up an environment which I am enjoying but who have also influenced my thoughts on many problems during almost daily arguments and joint work. Three friends and colleagues have given freely of their time by reading and criticizing the manuscript. Jeremy Knowles (Oxford), John Holbrook (Bristol) and David Trentham (Bristol) deserve much credit for a number of improvements and corrections, but they are, of course, not responsible for the inadequacies which remain. André Persoons and Karel Heremans (Leuven) have made a number of important corrections at the proof stage. For the efficient preparation of the manuscript, proof reading and the preparation of lists of references and the index I am indebted to Mrs. Janet Denton and Mrs. Mary Gutfreund.

Bristol,   H. GUTFREUND
August, 1971

Contents

Acknowledgements

The author wishes to acknowledge the co-operation of the following publishers for granting permission for the reproduction of diagrams from their publications:

Academic Press, Inc., New York

Figure 9 (Winlund and Chamberlain, *Biochem. Biophys. Res. Comm.* (1970), **40,** 43)
Figure 13 (Hess et al, *Biochem. Biophys. Res. Comm.* (1966), **24,** 824)

The American Society of Biological Chemists, Inc., Bethesda

Figure 3 (Bernhard, *J. biol. Chem.* (1956), **218,** 961)
Figure 6 (Hall et al, *J. biol. Chem.* (1969), **244,** 3391)
Figure 15 (Frieden and Colman, *J. biol. Chem.* (1967), **242,** 1705)
Figure 45 (Bright and Gibson, *J. biol. Chem.* (1967), **242,** 994)
Figure 52 (Heck, *J. biol. Chem.* (1969), **244,** 4375)

The Biochemical Journal, London

Figure 7 (Trentham et al, *Biochem. J.* (1969), **114,**19)
Figure 33 (Engel and Dalziel, *Biochem. J.* (1969), **115,** 621)
Figures 49, 50 and 51 (Stinson and Gutfreund, *Biochem. J.* (1971), **121,** 235)

European Journal of Biochemistry, Liège

Figure 14 (Cohen, *Europ. J. Biochem.* (1969), **11,** 520)

Federation of European Biochemical Societies, Liège

Figure 44 (Hess and Wurster, *FEBS Letters* (1970), **2,** 73)

Walter de Gruyter & Co., Berlin

Figure 43 (Wurster and Hess, *Hoppe—Seyler's Z.* (1970), **351,** 869)

International Union of Pure and Applied Chemistry, Oxford

Figure 20 (Diebler, *Pure and Applied Chem.* (1969), **20,** 93)

National Academy of Sciences, Washington

Figure 36 (Levy et al, *Proc. Nat. Acad. Sci (USA),* **45,** 785)
Figure 56 (Kirschner et al, *Proc. Nat. Acad. Sci (USA),* **56,** 1661)

CHAPTER 1

Forces and Structure in Aqueous Solutions

1.1. THE QUANTITATIVE INTERPRETATION OF PHYSICAL MEASUREMENTS

Some thought about units and dimensions, as well as numerical values for certain physical properties, is a good preparation for dealing with physical processes. During the course of this book we shall be dealing with 'fast' reactions, 'large' molecules and 'close' neighbours. We have to be numerate and define how fast is fast, how large is large, and so on. However, the most valuable attribute of the experienced research worker is not only to know how accurate his results and statements have to be, but also what approximations he can make to come to a useful decision. There is no point in a superficial discussion of scientific inference and the statistical design and interpretation of experiments. Wilson (1952) wrote an admirable book on various aspects of this problem and Moroney (1951) presents statistical methods in an entertaining form. In the type of work discussed in the present text one rarely if ever deals with a large number of observations or samples. They are usually of the order of 10 rather than 100. Under these circumstances it is unwise to use statistics as anything but a guide to further work.

The accuracy required to prove a point, decide between alternative hypotheses or detect a phenomenon determines the development of techniques. Many of the most fundamental advances in science are due to the improvement of the resolution of measurements. Generally one uses a technique which is just accurate enough to solve the problem in hand. It is of doubtful value to collect accurate information without a purpose in mind. The mere thought that someone will want the information to solve some as yet unknown problem is too random a process for the progress of science. The chances are that the information will be needed for conditions quite different from those used in the old experiment.

For experimental measurements as well as for theoretical calculations in a multi-disciplinary field such as biophysical chemistry it is particularly important to keep the problems in mind and not to let the experimental technique or the computer take one into a blind alley.

A thorough understanding of units and dimensions provides one with a tool for checking the equations set up to describe the physical behaviour of a system. Subsequent numerical calculations are, of course, entirely dependent

1

on the units used. Two systems of units are in common use: the MKSA system (meters, kilograms, seconds, amperes) favoured in England and the cgs system (centimeters, grams, seconds) still favoured in the United States literature. In addition to these systems there are in use the practical units. In the areas covered in this book the most commonly used practical units are calories (to interpret heat measurements) and electronvolts (to interpret electrical measurements). It is the aim of this section to help the reader to familiarize himself with various units used rather than to preach a sermon in favour of one convention or another.

Table 1 gives the dimensions and different units for the quantities used in this book. A brief discussion of the different units used for energy measured by different techniques will be used to illustrate a number of points. Energy or work (W) can be expressed by

$$W = \int_{r_1}^{r_2} F \, dr$$

which is the displacement of a particle over the distance $L = r_2 - r_1$ with the force F acting upon it. The dimensions of force are MLt^{-2}, which is derived from the definition of force in terms of the acceleration (Lt^{-2}) it produces on a unit of mass. Force in cgs units is expressed in dynes (1 dyne = g cm sec^{-2}), while in MKSA units it is expressed in newtons (1 newton = kg m sec^{-2})

$$1 \text{ newton} \equiv 10^5 \text{ dynes}$$

The dimensions of work (ML^2t^{-2}) are derived from those of force times L. In cgs units work is expressed in ergs (1 erg = g cm^2 sec^{-2}), while in MKSA units it is expressed in joules (1 joule = kg m^2 sec^{-2})

$$1 \text{ joule} \equiv 10^7 \text{ ergs}$$

As indicated above, calculations of energy changes derived from calorimetric measurements are often expressed in calories. A calorie is the amount of heat required to raise the temperature of 1 g of water by 1 degree. If a chemical reaction is carried out in a well-insulated vessel (adiabatically) the solution will warm up if the reaction involves a negative heat change (exothermic) and *vice versa*. With a small correction for the difference in specific heat between water and the solution, the heat change in calories is easily calculated from the temperature change:

$$1 \text{ calorie} \equiv 4 \cdot 186 \text{ joules}$$

The above discussion shows the relation between mechanical and thermal energy and now electrical energy will be introduced. Next to thermal measurements, electrical measurements are probably most frequently used

Table 1

	Dimensions	c.g.s. units	MKSA units	Practical units
Length	L	cm	m	$m\mu = 10^{-9}$ m
Area	L^2	cm^2	m^2	$A = 10^{-8}$ cm $= 10^{-10}$ m
Volume	L^3	cm^3	m^3	
Mass	M	g	kg	
Time	t	sec	sec	
Velocity	Lt^{-1}	cm sec^{-1}	m sec^{-1}	
Acceleration	Lt^{-2}	cm sec^{-2}	m sec^{-2}	
Force	MLt^{-2}	g cm sec^{-2} (dyne)	kg m sec^{-2} (newton)	newton $= 10^5$ dyne
Work Energy	ML^2t^{-2}	g cm^2 sec^{-2} (erg)	kg m^2 sec^{-2} (joule)	calorie $= 4 \cdot 186$ joules
Power	ML^2t^{-3}	g cm^2 sec^{-3}	kg m^2 sec^{-3}	watt $=$ joules sec^{-1}
Density	ML^{-3}	g cm^{-3}	kg m^{-3}	
Pressure	$ML^{-1}t^{-2}$	g cm^{-1} sec^{-2} (dyne cm^{-2})	kg m^{-1} sec^{-2}	Atmosphere $= 10^6$ dyne cm^{-2}
Electric current		(erg cm)$^{\frac{1}{2}}$ sec^{-1}	Ampere (A)	A $=$ coulomb sec^{-1}
Electric charge		(erg cm)$^{\frac{1}{2}}$	A sec $=$ coulomb	
Electric potential		(erg cm^{-1})$^{\frac{1}{2}}$	volt $=$ joules coulomb^{-1}	
Electrical energy		erg		(e.s.u.)^2 cm^{-1}

for the study of energy changes. The electrical measurements are commonly made in terms of volts and amperes. The MKSA system has amperes as a dimension but for practical purposes the following relations should be discussed:

$$\text{amperes} = (\text{erg cm})^{-\frac{1}{2}} \sec^{-1} = \text{coulomb sec}^{-1}$$

$$\text{volts} = \text{joules coulomb}^{-1}$$

The coulomb corresponds to 6.28×10^{18} charges (electrons or protons) and one mole equivalent of charges is 9.65×10^4 coulombs (1 Faraday). The number of particles in a mole equivalent is Avogadro's number

$$6.28 \times 10^{18} \times 9.65 \times 10^4 = 6.03 \times 10^{23}$$

The fractional exponential in the units for amperes given above comes from the relation between charge and force:

$$F = e_1 e_2/r^2*$$

The force between two charged particles is equal to the product of the number of charges (e_1 and e_2) on each, divided by the square of the distance between them (see Section 1.4.2). If force is expressed in the units g sec^{-2} cm it can be seen from the equation for F that the product of two charges has the units $\text{g sec}^{-2} \text{cm}^3$ or ergs cm. In consequence fractional exponentials occur for the units of single charges. This problem is avoided by the use of amperes in the MKSA system. The charge is defined as amp sec and consequently ampere is the rate at which current flows (coulomb $\sec^{-1}$).

Electrical power is expressed in watts = amps × volts. Therefore,

$$1 \text{ watt} = 1 \text{ joule sec}^{-1}$$

This relation is helpful for the electrical calibration of calorimetric equipment.

A unit which is frequently used to describe such phenomena as ionization and electrode potentials is the electronvolt (eV). One electronvolt is the energy gained by a charged particle when its potential is raised by 1 volt:

$$1 \text{ eV} = 1.59 \times 10^{-12} \text{ ergs} = 1.59 \times 10^{-19} \text{ joules}$$

and for a mole equivalent this becomes

$$1.59 \times 10^{-19} \times 6.03 \times 10^{23} = 9.65 \times 10^4 \text{ joules} = 23 \text{ kcal/mole}$$

The interaction of electromagnetic radiation with chemical systems results in energy changes and the relation between these and thermal, electrical and

---

* This form of Coulomb's law, used throughout this book, is for cgs units. For MKSA units $F = e_1 e_2/4\pi\varepsilon_0 r^2$, where $\varepsilon = \text{coulomb}^2/\text{newton} \times \text{meter}^2$ and $4\pi\varepsilon_0 = c^{-2} \times 10^7$ (c is the speed of light).

chemical energies can be illustrated by a simple example. The visible range of wavelength is between 400 and 700 mμ (mμ = nm = 10^{-9} m). The energy of electromagnetic radiation expressed per quantum or photon is

$$E = hv = hc/\lambda$$

where h is Planck's constant (6·626 × 10^{-27} erg sec), c is the speed of light (3 × 10^{10} cm sec^{-1}), v is the frequency corresponding to the wavelength λ. The energy of one mole equivalent, 6·03 × 10^{23}, of photons of wavelength 600 mμ (or 600 × 10^{-7} cm) is given by

$$E = \frac{6 \cdot 626 \times 10^{-27} \times 3 \times 10^{10}}{600 \times 10^{-7}} \times 6 \cdot 03 \times 10^{23}$$

$$= 2 \cdot 01 \times 10^{12} \text{ ergs} = 2 \cdot 01 \times 10^5 \text{ joules} = 48 \text{ kcal/mole}$$

This means that the maximum energy gained by absorption of light at 600 mμ is 48 kcal/mole. Substitution of different wavelength in the above calculation shows that light of shorter wavelength can produce larger energy changes.

An understanding of all aspects of the interconversion of different forms of energy is of importance both for the design and interpretation of physical measurements and for an understanding of biological processes in molecular terms. Many interesting phenomena involve the interconversion of chemical energy into mechanical energy, in muscle, into electrical energy, in nerve, or into light in bioluminescence. Conversely, the conversion of light energy into chemical energy is the key process in photosynthesis and acts as a control in other photobiological responses.

Table 2. Useful Numbers and Conversions

N Avogadro's number Number of molecules/mole $\}$	6·03 × 10^{23}
C Coulomb 6·28 × 10^{18} charges = 2·998 × 10^9 electrostatic units (e.s.u.)	
h Planck's constant 6·626 × 10^{-27} erg sec	
k Boltzmann's constant 1·381 × 10^{-16} erg deg^{-1}	
R Gas constant 1·987 cal deg^{-1} mole^{-1}	
Electronvolt = 23 × 10^3 calories/mole	
Calorie = 4·186 joules	
1 e.s.u. corresponds to 2·08 × 10^9 electronic charges	
1 coulomb = 6·28 × 10^{18} charges	
1 Faraday = 9·65 × 10^4 coulombs	

Tables 1 and 2 provide a summary of the statements and derivations given in this section and give numerical values for the constants used throughout the text.

1.2. SOME THERMODYNAMIC DEFINITIONS

1.2.1. Intensive and Extensive Properties

In the present text the formalisms of thermodynamic principles are only briefly stated as an *aide-mémoire*. For a systematic treatment of thermodynamics one of many excellent texts should be consulted (for instance, Hill, 1966 and 1968; Kauzmann, 1967). Katchalsky and Curran (1965) provide an introduction with particular application of thermodynamics to biophysical topics. The concepts mentioned in this section will be used and explained in greater detail in further applications.

Intensive properties such as temperature or pressure are independent of the size of the system. It is common practice during an operation of changing temperature to specify constant pressure. When a quantity is a function of more than one independent variable (intensive property) it is necessary to use partial derivatives. For instance, a change in volume dV due to change in temperature dT and change in pressure dP is expressed by

$$dV = \left(\frac{\partial V}{\partial T}\right)_P dT + \left(\frac{\partial V}{\partial P}\right)_T dP$$

A useful elementary treatment of partial differential equations is given by Klotz (1964).

Heat, energy, free energy and entropy are extensive properties of a system. This means that they are dependent on the size of the system and have to be expressed as per mole or per litre, etc. The extensive properties used in the present text are:

Total Energy	E	The symbol Δ is used to indicate a
Gibbs Free Energy	G	difference: ΔE is the energy change
Heat or Enthalpy	H	during a specified operation.
Entropy	S	

The physical significance of free energy and equilibria will be discussed in detail in Sections 1.3.1 to 1.3.4. At present we need to accept the definitions that two states of a system are in equilibrium if there is no free energy change in going from one state to the other under the specified conditions, $\Delta G = 0$. *If there is a spontaneous change from one state to the other the system will go towards the state of lower free energy, $\Delta G < 0$.*

The heat change ΔH is identical with the energy change when the operation is carried out at constant volume, a condition applicable to all but exceptional examples discussed here:

$$\Delta H = \Delta E + P\Delta V$$

When the heat of a reaction is measured in a calorimeter an increase in temperature indicates $\Delta H < 0$: the process is exothermic, the heat content of the system decreases during the reaction. In connection with calculation of electrostatic energy it was shown that attractive forces decrease the energy of a system.

The relation between heat and free energy

$$\Delta G = \Delta H - T\Delta S$$

introduces the entropy function S.

1.2.2. Entropy and Organization

Discussions of entropy are often shrouded either in algebraic formalism or in philosophical mystery. This has resulted in some difficulties in the general understanding of this important concept. It is necessary that the meaning of entropy should be properly understood for the subsequent use of thermodynamic principles for the explanation of complex phenomena in macromolecular interactions. This should be possible if a brief description of the physical phenomena due to the entropy of a system is followed by some discussion of the principles involved in the calculation of entropy changes due to changes in the properties of the molecules which make up the system. This very elementary treatment will be supplemented by examples of its use as occasion demands (see also further discussion Section 5.1).

An ideal gas is defined phenomenologically as a gas which behaves according to the ideal gas law:

$$PV = nRT$$

The physical condition for a gas to behave as an ideal gas is that there is no molecular interaction (no forces between molecules) and that the gas molecules can be considered to take up a negligible part of the volume. These conditions usually apply as the pressure becomes very low. Let us consider what would happen if a certain number of gas molecules are transferred from a volume of 1 m^3 to a volume of 10 m^3 under such ideal conditions. The gas molecules will expand spontaneously to fill the larger volume. This spontaneous process will result in a decrease in free energy of the gas and because of the absence of any forces between the molecules there is no heat change: $\Delta H = 0$ and hence

$$\Delta G = -T\Delta S$$

This shows that the decrease of free energy in this idealized process is entirely due to an increase in entropy. The spontaneous process results in an increase in entropy; molecules will distribute themselves in such a way that the entropy of a system is at a maximum.

The above merely describes the phenomenon that a free energy change in the absence of a heat change is due to the entropy change. Now, what physical condition of the gas molecules is changed when a larger total volume is available to them all? The only thing which has changed is that the molecules have a larger number of volume elements among which they can distribute themselves. Therefore, the molecules have a larger number of ways in which they can distribute themselves among these volume elements.

When a solid phase is in equilibrium with a liquid phase and heat is supplied to the system at essentially constant temperature the following relations hold:

$$\Delta G = 0 \text{ (system in equilibrium)}$$

$$\Delta S = \frac{\Delta H}{T}$$

The heat taken up at constant temperature is a measure of the increase in entropy due to transferring material from the solid to the liquid phase. An essential feature of this 'idealized' process is that it is carried out in a completely reversible manner. This can only be done when the heat is supplied so slowly that none of it is wasted by raising the temperature of the system. The increase in entropy is again related to the more random distribution of molecules in a liquid as compared with a solid. The relation between entropy and heat during changes in interaction will be discussed further in Section 1.5.3.

Similar reasoning applies to the entropy change of mixing two systems in the gaseous, liquid or solid state. In the absence of other forces the mixture will eventually have a random distribution of molecules from the two different systems. Boltzmann developed the idea of relating entropy with the number of ways of arranging a system. Entropy is an extensive property and the entropy S_A of system A and entropy S_B of system B results in entropy S_{AB} for the mixed system

$$S_{AB} = S_A + S_B$$

However, the number of ways of arranging the mixed system N_{AB} is related to the number of ways of arranging the two separate systems N_A and N_B, respectively, by

$$N_{AB} = N_A N_B$$

Boltzmann related the additivity of entropy to the number of states through the equation

$$S = k \ln N$$

where k is the Boltzmann constant.

In a perfect crystal at absolute zero temperature (when, theoretically, there is only one state) $N = 1$ and the entropy of the crystal is zero. As the temperature is raised more electronic, vibrational and translational states become populated and the entropy increases with the number of possible states. The calculation of the absolute entropy is a complex theoretical problem, but the calculation of entropy changes or the entropy due to some particular feature of a molecule is often quite simple and an example is given on page 19.

1.3. EQUILIBRIUM THERMODYNAMICS

1.3.1. Rates and Equilibria

When a marble rolls into an equilibrium position it appears to be at rest there. In contrast, the chemical equilibrium is considered in terms of a balance of movements. In the simple equilibrium

$$CH_3COOH \underset{k_{-1}}{\overset{k_1}{\rightleftharpoons}} CH_3COO^- + H^+$$

there is a continuous rapid dissociation of acetic acid into acetate and hydrogen ions and a reverse association of these ions into acid. The kinetic requirement for equilibrium is given by the law of mass action which relates the velocity of a chemical process to the concentration of the reactants through velocity constants:

$$\frac{dC_{CH_3COO^-}}{dt} = k_1 C_{CH_3COOH} - k_{-1} C_{CH_3COO^-} C_{H^+}$$

and

$$\frac{dC_{CH_3COOH}}{dt} = k_{-1} C_{CH_3COO^-} C_{H^+} - k_1 C_{CH_3COOH}$$

At equilibrium there is no net change in the concentrations of the reactants and hence

$$k_1 \bar{C}_{CH_3COOH} = k_{-1} \bar{C}_{CH_3COO^-} \bar{C}_{H^+}$$

where $\bar{C}$ represents equilibrium concentrations

$$K = \frac{k_1}{k_{-1}} = \frac{\bar{C}_{CH_3COO^-} \bar{C}_{H^+}}{\bar{C}_{CH_3COOH}}$$

The thermodynamic requirement for equilibrium is that the free energy of the system is at a minimum. The free energy of the individual molecules is, however, distributed according to Boltzmann's distribution law in the form

$$\frac{N_{\Delta G}}{N_0} = e^{-\Delta G/RT}$$

which gives the fraction of molecules with ΔG above the average free energy.

1.3.2. Chemical Potentials

In a system of several components one can define the partial molar free energy $\bar{G}_A$ or chemical potential μ_A of component A by

$$\bar{G}_A = \mu_A = \left(\frac{\partial G}{\partial n_A}\right)_{T,P,n_B\ldots} \tag{1.1}$$

at constant pressure, temperature and concentrations of components other than A; n_A stands for the number of moles of component A.

The change dG in the free energy of a system can be expressed as the sum over all its components

$$dG = \mu_A\, dn_A + \mu_B\, dn_B + \ldots \tag{1.2}$$

If the simple equilibrium process $A \rightleftharpoons B$ is now considered in terms of chemical potentials. It is helpful to write the progress of the reaction (dx) towards equilibrium as follows:

$$dx = -dn_A = +dn_B$$

the decrease in free energy during this process will be

$$-dG = -(\mu_A\, dx - \mu_B\, dx) \tag{1.3}$$

The algebraic condition for a minimum of the approach to equilibrium illustrated in Figure 1 is given by

$$dG/dx = 0$$

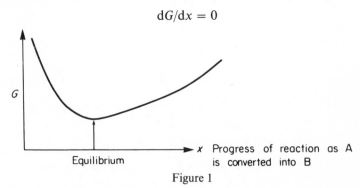

Figure 1

This means that at equilibrium

$$\mu_A = \mu_B$$

The condition that the chemical potentials of components of a system in equilibrium are equal is used to derive a number of important relations.

The ideal gas is again useful to illustrate a thermodynamic principle before applying this principle to real solutions of interest in biochemical reactions.

At normalized temperature and pressure (1 atm., 0° centigrade) 1 mole of gas takes up a volume of 22·42 l. of volume. This volume is directly proportional to changing temperature and inversely proportional to changing pressure according to the gas law

$$nRT = VP$$

where n is the number of moles of the gas. The molar volume of a gas at a given temperature and pressure is expressed as $\bar{V}$. The introduction of reference states ($P°$, the standard pressure of 1 atm., $G°$, the standard free energy and $\mu°$, the standard chemical potential of a gas under standard conditions) are necessary to provide integration constants for the solution of equations describing the system. The following relations (Maxwell's relations) also have to be introduced:

$$\left(\frac{\partial G}{\partial T}\right)_P = -S \quad \text{and} \quad \left(\frac{\partial G}{\partial P}\right)_T = V$$

Consequently

$$\left(\frac{\partial \mu}{\partial P}\right)_T = \bar{V}$$

and at constant temperature

$$d\mu = \bar{V}\,dP$$

the change $\int_{\mu°}^{\mu} d\mu = \mu - \mu°$ when the pressure changes $\int_{P°}^{P} dP = P - P°$ where P is the pressure of the gas with chemical potential μ^P

$$\mu^P - \mu° = \int_{P°}^{P} \frac{RT}{P}\,dP \qquad \text{(for one mole } \bar{V} = RT/P)$$

the integral $\int_{P°}^{P} (dP/P)RT$ is evaluated as $RT\ln P/P°$ if $P°$ is taken as 1

$$\mu = \mu° + RT\ln P \tag{1.4}$$

In an ideal solution the standard state of a solute is taken as $C_s° = 1$ molar and the standard chemical potential of the solute is $\mu_s°$. The chemical potential μ_s of a solute in an ideal solution at concentration C_s is

$$\mu_s = \mu° + RT\ln C_s \tag{1.5}$$

1.3.3. Activity and Concentration

So far all discussions have referred to ideal solutions and concentrations. Proportionalities of concentrations in thermodynamic equations are only valid in ideal gases and ideal solutions. Conversely, an ideal gas or ideal solution are defined as systems for which the gradient $(\partial G/\partial n_s)_T$ is independent of the absolute concentration. This is usually correct when the gas is at such low pressure or the solution so dilute that molecular interactions are negligible.

In real, or more concentrated solutions, three factors account for deviations from linear dependence of chemical potential on absolute concentration if defined as above:

(1) Solute–solute interaction due to intermolecular forces, which increases with increasing solute concentration.
(2) Solute–solvent interaction as, for instance, the solvation of ions.
(3) Volume taken up by the solute, which is considered as negligible (solute or gas molecules have been treated as points) in ideal systems.

A linear gradient is obtained if the chemical potential is defined in terms of activity instead of concentration. The activity of a solute a_s is related to the concentration through the activity coefficient f_s by

$$a_s = f_s C_s$$

and f_s approaches unity as C_s approaches zero. Although the term concentration will be used in further discussions it must be noted that expressions in this form are only strictly applicable to infinitely dilute solution.

1.3.4. Standard Free Energies and Equilibria

For the simple equilibrium

$$CH_3COOH \rightleftharpoons CH_3COO^- + H^+$$

$$AH \qquad\qquad A^- \qquad H^+$$

one can write the following expressions corresponding to those derived in the previous section

$$\mu_{AH} = \mu_{AH}^\circ + RT \ln C_{AH}$$

$$\mu_{A^-} = \mu_A^\circ + RT \ln C_{A^-}$$

$$\mu_{H^+} = \mu_{H^+}^\circ + RT \ln C_{H^+}$$

At equilibrium the sums of the chemical potentials on each side of the stoichiometric equation must be equal. Using the symbols $\bar\mu$ and $\bar C$ for potentials and concentrations at equilibrium the above can be rearranged as follows:

$$\bar\mu_{AH} = \bar\mu_{A^-} + \bar\mu_{H^+}$$

Therefore

$$\mu_{AH}^\circ - (\mu_A^\circ + \mu_{H^+}^\circ) = RT[\ln \bar C_{A^-} + \ln \bar C_{H^+} - \ln \bar C_{AH}]$$

also

$$\mu_{AH}^\circ - (\mu_A^\circ + \mu_{H^+}^\circ) = -\Delta G^\circ$$

The standard free energy change ΔG° of transforming AH into $A^- + H^+$ refers to a system with all reactants and products in the standard state, i.e. molar:

$$\Delta G^\circ = -RT \ln \frac{\bar{C}_A \cdot \bar{C}_{H^+}}{\bar{C}_{AH}} = -RT \ln K$$

It is important to distinguish between the standard free energy of a process with all components in their standard state and the free energy of a reaction at some specified condition. In Section 1.3.2 it has been shown that at equilibrium the free energy of transformation of reactant into product is zero. The free energy of a reaction at some specified concentrations C_{A^-}, C_{H^+} and C_{AH} can be derived from the above relations as follows:

$$\Delta G = \Delta G^\circ + RT \ln \frac{C_A \cdot C_{H^+}}{C_{AH}}$$

when

$$\frac{C_A \cdot C_{H^+}}{C_{AH}} = 1$$

then $\Delta G = \Delta G^\circ$ and when

$$RT \ln \frac{C_A \cdot C_{H^+}}{C_{AH}} = RT \ln K$$

then $\Delta G = 0$.

If numbers are introduced into the equation

$$\Delta G^\circ = -RT \ln K \text{ at } 25^\circ \text{ C (or } 298^\circ \text{ absolute) } RT = 591 \text{ cal mole}^{-1}$$

and after conversion into logarithms to base 10

$$\Delta G^\circ = -1.365 \log K \text{ in kcal}$$

This calculation shows that a modification of a reactant which results in a change of 1·365 kcal in the standard free energy of the reaction, results in a change of an order of magnitude in the equilibrium constant.

A tenfold change in the equilibrium concentrations of reactants, or of an equilibrium distribution among isomeric forms of a macromolecule, can often make the difference between one or other component being experimentally detectable at all. The relatively small amount of energy required to cause a large shift in equilibrium position is the basis for the comments on the very fine balance of forces in complex molecules in Sections 1.4.1 and 1.5.3.

1.3.5. Heats of Ionization (Standard Enthalpy)

The heat of a reaction can be determined either directly in a calorimeter or it can be calculated from the temperature dependence of the equilibrium. The temperature dependence of ionization equilibria has been studied extensively for the characterization of different groups. From the Maxwell relation

$$\left(\frac{\partial \Delta G}{\partial T}\right)_P = -\Delta S \tag{1.6}$$

it follows that

$$\Delta G = \Delta H + T\left(\frac{\partial \Delta G}{\partial T}\right)_P \tag{1.7}$$

differentiating with respect to T at constant P and rearrangement gives the Gibbs–Helmholtz equation

$$\left[\frac{\partial(\Delta G/T)}{\partial(1/T)}\right]_P = \Delta H \tag{1.8}$$

If $\Delta G°$, the standard free energy, is used $\Delta H°$, the standard enthalpy of the reaction is obtained. Substituting $-RT \ln K$ for $\Delta G°$ (see Section 1.2.4) we obtain

$$\left(\frac{\partial \ln K}{\partial(1/T)}\right)_P = -\frac{\Delta H°}{R} \tag{1.9}$$

If $\Delta H°$ is independent of temperature, K_P is the equilibrium constant at constant pressure

$$\ln K_P = \frac{-\Delta H°}{RT} + C$$

and a plot of ln K against $1/T$ has the slope of $-\Delta H°/R$. The heat of a reaction from equilibrium measurements at two temperatures can be calculated from van't Hoff's reaction isochore (with conversion to decadic logarithms):

$$\log \frac{K_2}{K_1} = \frac{\Delta H°(T_2 - T_1)}{2 \cdot 303\, R T_1 T_2} \tag{1.10}$$

The van't Hoff equation is also very useful for the calculation of the change in equilibrium constant with temperature if the heat of the reaction is known. An example of this latter application is a calculation of the change in pH with temperature discussed in Section 2.2.1.

1.4. SOME PROPERTIES OF WATER

1.4.1. Hydrogen Bonding

All biological processes are either directly or indirectly under the influence of some of the characteristic properties of water. The special properties of water are due to the fact that it retains a relatively ordered structure as a liquid. This in turn is related to the structure of individual water molecules which results in strong dipolar interaction. A water molecule interacts with others in the following way:

The electron distribution in the H—O bonds is such that the oxygen atom is electronegative and a 'hydrogen bond' is formed with the oxygen atoms of two neighbouring water molecules sharing a hydrogen atom. An oxygen–oxygen distance of 2·76 A occurs frequently in liquid water. Hydrogen bonds are of considerable importance in the maintenance of the structure of large molecules and in the interaction of small molecules with large ones. Such bonding occurs when the steric requirements are fulfilled so that two electronegative systems can share a hydrogen nucleus. In some cases the interaction is ionic:

In other cases two dipolar structures are involved:

The electron distribution in $C=O$ and $C-N$ bonds is such that dipoles are formed similar to those found in the H_2O molecule. Water can form hydrogen bonds with material dissolved in it and consequently there is a competition between the intramolecular hydrogen bonds stabilizing the structure of the solute and solute–solvent bonding. During further discussions of macromolecular structure and protein–ligand interaction in water, it will become more and more apparent that there is a very fine balance of forces which is critical for these specific phenomena.

1.4.2. Dielectric Constant and Electrostatic Forces

The interaction of electric charges depends on the dielectric constant of the medium between them. The dielectric constant D of a medium can be defined by the force between two charged particles with e_1 and e_2 number of electronic charges, respectively, suspended in this medium at a distance r

$$F = e_1 e_2 / r^2 D \qquad (1.11)$$

For this purpose one assumes point charges, which means that the distance r must be large compared with the size of the particles.

The force is negative for the attraction between two oppositely charged particles and positive for the repulsion of like charged particles. It will be seen below that this is taken into account for the convention of signs for energy and heat changes during chemical processes.

The work W which has to be performed, or energy to be expended on a system to separate two charges interacting with force F from distance r_1 to r_2 is given by the integration

$$W = \int_{r_1}^{r_2} F \, dr$$

and the energy expended to keep two particles at a distance r is from (1.11) and (1.12)

$$W = e_1 e_2 / r D$$

The dielectric constant in a vacuum is 1, while the dielectric constant of water is 80. This large dielectric constant has the following interesting consequence: charge interactions in a purely aqueous medium are relatively weak. For example, the energy of interaction between a Na^+ ion and a Cl^- at their distance in a crystal (2·81 A) is for a single pair:

$$\frac{2·3068 \times 10^{-19} \text{ erg cm}}{2·81 \times 10^{-8} \text{ cm}} = 0·82 \times 10^{-11} \text{ erg}$$

For a mole equivalent of pairs this comes to

$$0·82 \times 10^{-11} \times 6·03 \times 10^{23} = 4·96 \times 10^{12} \text{ erg/mole}$$

which is equivalent to

$$\left. \begin{array}{c} 4·96 \times 10^5 \text{ joules} \\ 5·18 \text{ eV} \\ 118 \text{ kcal} \end{array} \right\} \text{ per mole}$$

The heat of evaporation and dissociation into ions for sodium chloride crystals is 183 kcal/mole. The above calculation only takes into account the forces between an ion and one of its neighbours of opposite charge. The stability of the crystal also involves interactions with other ions in the vicinity.

If we now consider the interaction of two ions at a distance of 2·81 A in a medium of dielectric constant 80, we note that the interaction energy is reduced from 118 kcal to about 1·5 kcal. This effect of the dielectric constant of water, together with the interaction between ions and water to form larger solvated ions, results in their dissociation.

A comparison of the dielectric constants of some liquids is made in Table 3. We find that the addition of CH_2 units decreases the dielectric constant of the resulting liquid and the removal of the polar OH group causes a further

Table 3

Compound	Dielectric constant at 25° C
Water	78·54
Methanol	32·63
Ethanol	24·30
Hexanol	13·30
Cyclohexane	2·22

decrease. Attempts to calculate the interaction energies of charged groups in a complex system such as a protein solution are confronted with the following difficulties. The charges affect the local dielectric constant and so does the environment of the charges. A protein molecule can offer a great variety of environments depending on the particular amino acid residues in the vicinity of the charges which are being considered. In this way the hydrocarbon chains of amino acid residues have very specific effects and the macromolecule has the facility to control interactions through changes in the environment induced by conformation changes.

In addition to the changes in the local dielectric constant, the environment of hydrocarbon residues can also stabilize intramolecular hydrogen bonds by excluding water molecules from competition for such bonding. The role of such 'hydrophobic' areas in the reactions of protein molecules will be discussed further in Section 1.5.1.

1.4.3. Protons and the Hydration of Ions

In pure water the concentrations of H^+ and OH^- ions are equal, each is 10^{-7} M. This statement needs some expansion since these ions do not exist strictly in this form in water. The proton (H^+) is always combined with a water molecule to form a hydronium ion (H_3O^+) and this in turn is hydrated by a hydrogen-bonded shell of three water molecules. Consequently, the following structure results from a nominal hydrogen ion in solution, which is $H_9O_4^+$

$$
\begin{array}{ccc}
H & & H \\
\diagdown & & \diagup \\
& O & \\
& \vdots & \\
& H & \\
& | \; + & \\
& O & \\
\diagup & & \diagdown \\
H & & H \\
\diagup & & \diagdown \\
H{-}O & & O{-}H \\
| & & | \\
H & & H
\end{array}
$$

The hydration of ions deserves some special comment since it is responsible for many of their characteristic properties. If we consider a series of monovalent ions with progressively larger ionic radii, as measured in crystal lattices: Li^+, Na^+, K^+, we find that their sizes in solution, judged from their mobility in water, are in inverse order. This is due to the larger charge density around a small ion resulting in stronger hydration.

If one is concerned with mechanisms controlled by the presence of one of these ions, for example transport through membranes, specific binding resulting in conformation changes or control of catalysis, one often wishes to interpret specificity in relation to the size of the ion. In such cases one also has to consider the possibility that the specific binding or transport process could involve the stripping of the hydration layer from the ion. The replacement of the solvation layer by hydrophilic groups of specific carriers has been postulated for some transport mechanisms.

The ionic mobility in an electric field can be used to calculate the radius of ions (see Section 5.4.3). The larger the solvated ion the slower it migrates. The comparison of mobilities given here shows that protons move very fast:

Ion	Mobility in $cm^2\,volt^{-1}\,sec^{-1}$
Li^+	$40 \cdot 1 \times 10^{-5}$
Na^+	$52 \cdot 0 \times 10^{-5}$
K^+	$76 \cdot 2 \times 10^{-5}$
H^+	$363 \cdot 0 \times 10^{-5}$

There are two reasons why the mobility of protons is out of step with the series of ions which migrate faster the larger the unsolvated ionic radius.

Firstly the ion to be considered from the point of view of size is H_3O^+ and not H^+, and secondly hydrogen ions move in water in a special way. Inspection of the diagram of the hydrated hydronium ion suggests that protons can jump between two possible positions between two oxygen atoms:

$$\begin{array}{ccc} \diagdown & \diagup & \diagdown \\ O\cdots H-O & \rightarrow & O-H\cdots O \\ \diagup & \diagdown & \diagup \end{array}$$

In this way the movement of protons can be propagated by a rapid game of musical chairs and not by migration of a particular proton.

1.4.4. Ice and Water

In crystalline water, ice, the system of hydrogen bonds between water molecules is even more stable than in the liquid state and the rate of proton movement from one molecule to another is even faster than in the liquid. In fact, this is one of the fastest chemical processes investigated (see Section 7.1.3). Although ice itself is not found in biological systems, water is immobilized in some ordered form in solvation layers and in cavities surrounded by hydrophobic material. The properties of these solvent layers in contact with reactive areas of macromolecules are of great importance in determining the reactivity of groups within this area. This is a very active field of current research.

Studies of the transition of ice to water and water vapour give useful information about the amount of organization in water. The heat of sublimation of ice (12·2 kcal/mole) gives the heat required to separate water molecules from their positions in the ice crystal to that in water vapour. The van der Waals' attraction between water molecules is estimated at about 3 kcal/mole, which leaves 9 kcal for breaking two hydrogen bonds. The fusion of ice, which gives the heat change of going from the crystalline to the liquid state, requires 1·44 kcal/mole. This indicates that only 16 per cent. of the hydrogen bonds are broken during the melting of ice and that water is really a very ordered liquid.

An example of the calculation of the entropy of a selected phenomenon is the interpretation of the residual entropy in ice due to the mobility of protons referred to above. In a mole of ice there are $2N$ hydrogen nuclei and N oxygen atoms. If each hydrogen nucleus had a free choice to be in either covalent or hydrogen-bonded position there would be 2^{2N} possible configurations in a mole of ice. Let us add the condition that each oxygen molecule can only have two covalent bonds with hydrogen atoms, that means that only H_2O molecules are formed. There are 2^4 ways of arranging 4H atoms on one O with either hydrogen or covalent bonding, but only six of these produce H_2O. Hence, only 6/16 of all the 2^{2N} states are possible for N oxygen atoms:

$$2^{2N} \times (\tfrac{3}{8})^N = (2 \times 2 \times \tfrac{3}{8})^N = (\tfrac{3}{2})^N$$

The entropy due to the proton jumps in ice S_P is then

$$S_P = k \ln (\tfrac{3}{2})^N = 0.806 \text{ cal/mole/degree}$$

The experimental measurement of the residual entropy of ice crystals at low temperature gives the value 0.82 cal/mole/degree. This provides good support for the validity of the proposed model of disorder in ice crystals.

The procedure of using statistical methods for evaluating the number of states or configurations of some specific property has found wide application in the theoretical interpretation of the thermodynamic behaviour of large molecules.

1.4.5. Physiological Consequences of the Properties of Water

The dipolar structure of the water molecule and the consequent strong interaction of these molecules in the liquid phase are responsible for what L. J. Henderson (1913) calls 'The Fitness of the Environment'.

In the above sections we have been concerned with the variety of ways in which the properties of water affect the structure and reactivity of molecules dissolved in it. This will be a recurrent topic during further discussions presented here. There is one physiological mechanism which uses two macroscopic properties of water: its large specific heat and its large heat of evaporation. These two properties make water ideally suitable to keep organisms at a constant temperature while heat-producing chemical reactions are proceeding within them.

The specific heat of a substance is defined as the amount of heat required to heat 1 g of a substance by 1° C at 15° C. The heat is measured in calories. If we measure the heat required to heat 1 kg of a substance by 1° C it is expressed in kilocalories. This indicates that the larger the specific heat of a substance the better a thermostat it will make. The calorie is so defined that the specific heat of water is 1 and this is only surpassed among liquids by ammonia.

The heat of evaporation of water (12 kcal/mole) is again very much larger than that of any other common liquid. It requires 0.58 kcal to evaporate 1 g of water—the cooling effect of water evaporating on one's skin is well known. This provides an ideal mechanism for the dissipation of 'waste' heat from metabolic processes through perspiration.

The term waste heat should be used cautiously. Warm-blooded animals require a sophisticated control for temperature regulation, which must involve the production of heat as well as its removal.

1.5. HYDROPHOBIC INTERACTIONS

1.5.1. Solubility

It is common experience that ionic and polar molecules dissolve freely in water while non-polar substances such as benzene or long-chain hydrocarbons are not soluble in water. Solubility, like the extent of any process, is dependent upon the free energy change during the process. A sparingly soluble substance has a large positive free energy of solution. The solubility of a substance in water depends on whether it is energetically more favourable (lower free energy) when solute molecules are in contact with other solute molecules or when they are in contact with water. If they prefer interaction with water they are soluble, while preference for contact among themselves keeps solute molecules out of solution.

Covalent or electrostatic interactions between atoms or molecules, which were referred to in earlier sections, are conceptually somewhat different from hydrophobic interactions to be discussed now. A somewhat simplified explanation is that putting a hydrocarbon molecule into water is equivalent to making a hole in the water structure. The increase in free energy is dependent on the surface area of the hole. The tendency towards the lowest possible free energy will give a tendency towards reducing the surface area. When the hydrocarbon molecules stick together the surface area is reduced. This analogy of the energetic explanation for molecules which do not form dipolar or ionic bonds with water to stick together is, like most analogies, an over simplification which does not explain all the facts. The thermodynamic properties of water at an air–water interface are not the same as those of water in a shell around hydrophobic molecules. It is apparent that water can exist in a number of different highly ordered states in addition to ice and pure liquid. These different forms of water at air or hydrocarbon interfaces have not yet been fully elucidated. This problem is being investigated by comparisons of theoretical predictions from model structures with experimental measurements of thermodynamic and other physical properties. A detailed knowledge of the properties of 'local' water is important for predictions of reactivities of groups within this special solvent. An interesting difference between the transition from water to ice and from free water to water of solvation is in the respective volume change. During the formation of ice crystals the molar volume of H_2O increases, while the dissolution of non-polar or polar molecules in water results in a decrease in volume.

The difference in solubility of a solute in two different solvents gives a measure of the difference in the free energy of solution and hence the free energy change of transferring the solute from one solvent to another. Procedures for resolving the parameters ΔG, ΔH and ΔS from equilibrium measurements were briefly discussed in Section 1.3.5. It is also possible to measure ΔH directly in a calorimeter. Table 4 gives the thermodynamic parameters for the transfer of hydrocarbons from organic solvents to water.

Table 4 (from Kauzmann, 1959)

Transfer process	$T\Delta S$ transfer (kcal/mole)	ΔH transfer (kcal/mole)
CH_4 in $C_6H_6 \rightarrow CH_4$ in H_2O	−5·4	−2·8
n-Butane liquid $\rightarrow$ n-Butane in H_2O	−6·9	−1·0
C_6H_6 liquid $\rightarrow$ C_6H_6 in H_2O	−4·2	0
$C_6H_5C_2H_5$ liquid $\rightarrow$ $C_6H_5C_2H_5$ in H_2O	−5·7	0

Before attempting to analyse these thermodynamic data in terms of mechanisms or models, some attention should be paid to solubility properties of amino acids. These are of specific interest for the interpretation of the energy changes during intra- and intermolecular interactions of protein molecules.

The versatility of the side-chain structure of amino acids, which are polymerised through peptide bonds to form the primary structure of proteins, lies partly in the reactivity of ionic or nucleophilic groups and partly in specific structures of hydrocarbon residues. When one compares the solubilities of amino acids in different solvents and at different temperatures one can separate contributions from these different functions to the total solubility. The data in Table 5 show that the substitution of $-CH_2CH(CH_3)_2$ in place of hydrogen at the α-carbon of glycine causes a decrease in the temperature coefficient. However, substitution by $-CH_2CH_2COO^-$ causes

Table 5. Solubility in water in g per 100 g of water: S_0 and S_{100} at 0° and 100°

Substance	S_0	S_{100}	S_{100}/S_0
Glycine	14·2	67·2	4·73
Leucine	3·8	8·3	2·14
Glutamic acid	0·34	14·0	41·17

a large increase in the temperature coefficient of solubility. This means that the non-polar side chain has a negative effect on the temperature dependence and that the solubility of the non-polar residues themselves would decrease with temperature. This is in agreement with the decrease in entropy due to transfer of hydrocarbon into water (see Table 4). When ΔH is nearly zero, $\Delta G = -T\Delta S$ and an increase in free energy of transfer into water with increase in temperature is due to an increase in entropy. As will be seen, this relation between temperature and solubility has the effect of making it thermodynamically more favourable for hydrophobic residues to stick together in an aqueous environment as the temperature is raised. Ionic interactions, on the other hand, may remain nearly constant and consequently relatively less important.

1.5.2. A Plausible Model for Hydrophobic Interactions

So far we have discussed the thermodynamic consequences of putting non-polar molecules or parts of molecules into an aqueous environment. It now remains to interpret these phenomena in terms of molecular events and to show how one can interpret the structure and interactions of protein molecules in terms of the phenomena reported above.

A number of reviews (for instance, Kauzmann 1959, Scheraga 1963 and Jencks, 1969) give the background to investigations and proposed models. While only a single model is discussed here it must be emphasized that it is a good one rather than a complete one.

Scheraga (1963) has proposed the following explanation for the decrease in entropy—increase in order—when hydrocarbon residues are dissolved in water. As mentioned above, there is a considerable amount of local structure in water due to the fact that each water molecule can form up to four hydrogen bonds:

Due to this interaction a water molecule can be, at any one instant, in one of five states forming 0, 1, 2, 3 or 4 hydrogen bonds, respectively. It is proposed that the principle effect of insertion of a hydrocarbon molecule into water is on a single layer of water molecules around the solute molecule. This layer will consist of water molecules in the 0, 1, 2, 3 hydrogen-bonded state. Water molecules in the 4 hydrogen-bonded state will be relatively unaffected by the insertion of hydrocarbon molecules. The molecules in the layer around the hydrocarbon molecules, which have some of their hydrogen bonds replaced by contact with hydrocarbon, will have raised energy levels. Figure 2 shows the change in energy levels schematically. There is some suggestion that the energy of any completely hydrogen-bonded water molecule in the solvation layer would be lowered. We have seen that the principal thermodynamic consequence of this formation of a layer of water around a hydrocarbon molecule is a decrease in entropy. The heat change due to the breaking of hydrogen bonds ($\Delta H°$ is negative but small) is more than compensated

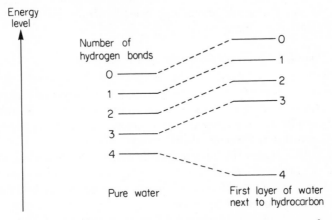

Figure 2   Changes of energy levels of water molecules on transfer from pure water to the first water layer next to a hydrocarbon solute (see, for instance, Scheraga, 1963)

by the entropy change ($\Delta S°$ is negative) to give a positive free energy change on addition $\Delta G° = \Delta H° - T\Delta S°$.

The strong remaining hydrogen bonds in the layer around hydrophobic residues must create a water structure which is more ordered than that of pure water.

1.5.3. Consequences of Hydrophobic Interactions

The thermodynamic parameters of hydrophobic groups interacting with each other are of opposite sign from those of hydrophobic groups interacting with water. As hydrophobic groups combine, the ordered water layers are liberated into the state of pure water and there is an increase in entropy and decrease in free energy. The magnitude of the favourable free energy change for the combination of hydrophobic groups is dependent on their size and on how well they fit together sterically. The better the specific fit of two combining groups the more water will be excluded from hydrophobic surfaces. This latter phenomenon makes hydrophobic interaction ideally suited for very specific biological complex formation.

When the first two or three complete structures of protein molecules became available, the generalization was deduced that most hydrophobic amino acid residues are inside the folded molecule while most polar groups are on the outside in contact with water. This would provide an energy minimum for a structure, not only because of the favourable contacts between the hydrocarbon residues but also because polar groups have a lower energy in water compared with being embedded in a non-polar medium. As more three-dimensional structures are being elucidated the truth of and the interesting exceptions from this rule will be examined (Klotz, 1970).

Another consequence of the formation of hydrophobic regions in protein molecules is the strengthening of specific hydrogen and ionic bonds. It has been mentioned in Section 1.4.1 that the formation of hydrogen bonds within or between solute molecules in aqueous solution occurs in competition with bonding between the groups on the solute molecule and water. If the intramolecular hydrogen bond is in a non-polar environment, water is excluded from the competition and the decrease in the local dielectric constant increases the ionic character and energy of the bond.

The stability of the native structure of a protein molecule depends on a complex interrelation of different forces. This results in the great sensitivity of the equilibrium between native and denatured protein to small changes in conditions and environment. An increase in temperature will stabilize some bonds and weaken others. Similarly, addition of non-polar solvents will weaken hydrophobic and strengthen ionic interactions. Any change from some optimum is likely to upset the balance of forces unless one or other type is overwhelmingly responsible for the correct structure. In relatively simpler processes, such as protein–protein interaction, there are situations where the role of one type of bonding predominates (see below).

As with all generalizations the exceptions are very interesting. The surface of a protein molecule often has hydrophobic areas for specific combining sites. The catalytic sites of enzymes have ionic and other polar groups to promote catalysis. The combination of hydrophobic combining sites with polar catalytic sites results in the formation of hydrophobic pockets with the polar groups in a highly reactive state.

The protein molecule, which evolves for the catalysis of a metabolic reaction or for some other function within the living organism, has to be suitable for survival and organization within its environment, as well as for its specific purpose. The formation of protein molecules which are built up from a fixed number of sub-units and the interaction of different proteins to form multifunctional complexes are regulated to a considerable extent by hydrophobic interactions and by the effect of hydrophobic areas on the energies of other non-covalent bonding.

An often-quoted example of a case where the organization of protein molecules into a structure results in an *increase* in the entropy of the system is the crystallization of tobacco mosaic virus. When the virus is crystallized the solution cools down $\Delta H > 0$ (see Section 1.2.1). With the crystals in equilibrium with the solution $\Delta H = T\Delta S$ and an increase in entropy of crystallization is indicated. This increase in entropy is likely to be due to exclusion of water of solvation from the virus surface. Stauffer et al. (1970) review a number of thermodynamic investigations into the polymerisation of tobacco mosaic virus protein in aqueous solutions. Positive ΔH and ΔS values are obtained for this process.

In a number of other systems involving polymerisation of protein molecules it is also found that the heat of aggregation is positive, again indicating an increase in entropy during a process which appears to involve a higher state of organization. The explanation is the same as that given for the positive entropy of crystallization: the liberation of free water.

An interesting practical consequence is that cooling drives the equilibrium

$$\text{monomer} \rightleftharpoons \text{polymer} \qquad \Delta H > 0$$

to the left. If a system increases its heat content during a reaction it is promoted further by the supply of heat and *vice versa*. The depolymerisation of some enzyme systems on cooling is likely to be one of the causes of the cold lability of certain enzymes. The generalization that biological material is more stable when kept in a refrigerator has its interesting exceptions. Many enzymes are found in polymeric forms with two, four or more identical sub-units firmly but non-covalently linked together. Any dissociation of these quarternary structures usually results in great lability of the enzymes. The most likely cause of this loss of structural stability on dissociation is the approach of water to areas which are not available to water in the polymeric state.

CHAPTER 2

Ionic Equilibria and Reactivities

2.1. CLASSIFICATION

2.1.1. Acids and Bases

In this chapter the simple processes involving the equilibria of protons and electrons between donors and acceptors will be used to illustrate some of the fundamental laws governing equilibria. The use of equilibria to obtain thermodynamic data can also be demonstrated through a detailed study of ionization and oxidation-reduction. A brief description of the types of chemical reactions to be considered under these headings will be followed by a detailed discussion of some examples of interest for the interpretation of reactions of proteins.

For catalytic purposes, as distinct from specific binding processes, the amino-acid side chains of proteins may have a limited number of related functions as general acids, general bases, electrophiles or nucleophiles. Electron or hydride transfer processes appear to require some cofactor apart from the purely amino-acid make up of the apoprotein. It will be very interesting to see whether some of the suggestions that certain oxidation reactions (lipoxygenase, for instance) are catalysed by a protein free from cofactors will be substantiated.

The simplest definition of acids and bases, due to Brönsted, also turns out to be quite adequate and convenient for most groups encountered in enzyme catalysis (metal atoms bound to the active site are a special case). An acid is defined as a proton donor either in the form

$$AH \rightarrow A^- + H^+ \text{ when AH is a neutral acid}$$

or in the form

$$BH^+ \rightarrow B + H^+ \text{ when } BH^+ \text{ is a cationic acid.}$$

A base is defined as a proton acceptor and both A^- and B are bases. The 'strength' of an acid or base is judged by its dissociation constant. The term acid dissociation constant

$$K_A = \frac{C_{A^-}C_{H^+}}{C_{AH}} \quad \text{or} \quad K_A = \frac{C_B C_{H^+}}{C_{BH^+}}$$

is now generally used regardless of whether one is interested in the basic properties of A^- or B or the acid strength of AH or BH^+.

The term weak acid or weak base is fairly arbitrarily given to acids or bases with K_A values in the range 1 to 10^{-14} M. The reason for this will become apparent when the measurement of dissociation constants and properties of buffers are discussed.

The term basic dissociation constant K_B is now rarely used, but it is necessary to define it since it occurs in some of the classical studies of the ionic properties of amino acids. This is best done with two specific examples. From the above definition of acids and bases it follows that CH_3COOH and $CH_3NH_3^+$ are acids, while CH_3COO^- and CH_3NH_2 are bases. The dissociation constants K_A represent the following equilibria:

$$K_A = \frac{C_{CH_3COO^-} C_{H^+}}{C_{CH_3COOH}} = 1 \cdot 78 \times 10^{-5} \text{ M}$$

and

$$K_A = \frac{C_{CH_3NH_2} C_{H^+}}{C_{CH_3NH_3^+}} = 2 \times 10^{-11}$$

Let us now consider the consecutive equilibria for the second case

$$CH_3NH_2 + H_2O \rightleftharpoons CH_3NH_3OH \rightleftharpoons CH_3NH_3^+ + OH^-$$

CH_3NH_2 and CH_3NH_3OH are experimentally indistinguishable and we may write for the basic dissociation constant

$$K_B = \frac{C_{CH_3NH_3^+} C_{OH^-}}{C_{CH_3NH_3OH}} = 5 \times 10^{-4} \text{ M}$$

The product $C_{OH^-} C_{H^+}$ is 10^{-14} and hence

$$K_B = 10^{-14}/K_A$$

2.1.2. Oxidants and Reductants

The chemical classification of the oxidation-reduction processes of enzyme reactions is much more complex and less complete than that of acid-base equilibria.

Oxidation-reduction equilibria involve electron transfer between donor and acceptor. In some such reactions only electrons are transferred while in others a proton is transferred with the electron, resulting in hydrogen transfer. From the point of view of those interested in biological reactions the most important problems lie in explorations of compounds which act as a link between a series of steps involving electron transport and a series of steps involving hydrogen transport. There is considerable uncertainty about the sequence of events during hydrogen transfer; this problem will become more apparent below.

Oxidation is defined as a process involving the loss of an electron and reduction the uptake of an electron. If the rusting of iron is written as

$$4Fe + 3O_2 \rightarrow 4Fe^{3+} + 6O^{2-}(2Fe_2O_3)$$

one sees that iron is raised to the $+3$ oxidation state and oxygen is lowered to the -2 oxidation state. Similarly, a compound which has a higher affinity for electrons than ferrous iron can oxidize it to ferric iron

$$Fe^{2+} \rightarrow Fe^{3+} + e^-$$

Copper and some other metals can also act as direct electron donors or acceptors in biological systems.

Hydrogen transfer reactions occur in some cases through the transfer of one hydride ion and the liberation of a proton:

Catalysed by
alcohol dehydrogenase

while in other cases two hydrogen atoms are transferred by one of several possible mechanisms:

In addition to hydrogen transfer from, for instance, xanthine to O_2 via reduced flavine nucleotide to form uric acid and H_2O_2, flavoproteins can catalyse reactions of the type:

$$\left.\begin{array}{l} \text{Reduced flavoprotein} + 2Fe^{3+}\text{-porphyrin protein} \\[6pt] \text{Oxidized flavoprotein} + 2H^+ + 2Fe^{2+}\text{-porphyrin protein} \end{array}\right\}$$

Many of the environmental effects on proton transfer are of importance in controlling the rates and equilibria of electron transfer. The thermodynamic analysis of oxidation-reduction equilibria will be discussed in Sections 2.3.1 and 2.3.2.

2.2. PROPERTIES OF WEAK ACIDS AND BASES

2.2.1. Buffer Systems and pK

The simplest definition of weak acids and weak bases given on p. 28 will now be related to the experimental investigation of ionizing groups through the measurement of pH.

In discussions of the relation between C_{H^+}, C_{A^-} and C_{AH} the terms pH and pK are generally used:

$$\text{pH} = -\log C_{H^+} \quad \text{and} \quad \text{pK} = -\log K = -\log \frac{C_{A^-} \cdot C_{H^+}}{C_{AH}}$$

The relation between pH and pK is derived from the Henderson–Hasselbalch equation:

$$\log K = \log \frac{C_{A^-}}{C_{AH}} + \log C_{H^+}$$

$$\text{pK} = \text{pH} - \log \frac{C_{A^-}}{C_{AH}} \tag{2.1}$$

This equation shows that for a compound with a single ionizing group the acidic and basic forms of the compound will be at the same concentration when pH = pK. The form of the titration curve, relating pH to C_{A^-}/C_{AH}, shown in Figure 3, indicates that in the region pH = pK $\pm$ 0.5 of a weak acid the pH is relatively insensitive to the addition or removal of hydrogen ions. This phenomenon is called buffering and a weak acid near its pK is called a buffer.

Some buffers are relatively insensitive to changes in temperature (phosphate buffer for instance) while others change their pK, and with that the pH of the solution, when the temperature changes. These differences are due to the heat of ionization of the particular acid. A commonly used buffer for enzyme studies in solutions near pH 8 is trishydroxymethylaminomethane

$$C(CH_2OH)_3NH_3^+ \rightleftharpoons C(CH_2OH)_3NH_2 + H^+$$

This system, commonly referred to as tris-buffer, has a heat of ionization $\Delta H = +10.9$ kcal for the reaction in the direction shown above. From equation (1.10) one can derive

$$pK_1 - pK_2 = \frac{\Delta H}{4.6} \frac{T_2 - T_1}{T_1 T_2} \tag{2.2}$$

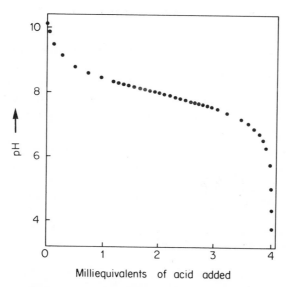

Figure 3 This titration curve of tris(trishydroxy-methylaminomethane) is taken from Bernhard (1956). 40 ml of a solution containing 4 milliequivalents of tris was titrated with 1.0 N HCl at 25°. The solution was 0.6 M with respect to KCl

for the change pK_1 to pK_2 on changing the temperature from T_1 to T_2. From this one can calculate that a tris-buffer solution adjusted to pH 8.0 at 25° C will change to pH 8.67 when cooled to 0° C (273° K). The positive ΔH signifies that raising the temperature will result in a shift of the equilibrium to the right-hand side of the above reaction, resulting in a liberation of hydrogen ions and a decrease in pH (see also Section 1.3.5 and Bernhard, 1956).

For the above calculation, as well as the discussion in Section 1.3.5, it was assumed that ΔH is constant over the temperature range considered.

This is not always correct. Changes in the structure of the reactants with temperature result in a change in heat capacity and a change in the slope $\ln K/(1/T)$. This point is discussed further in Section 7.2.3. The heat of ionization of carboxylic acids is very small and acetate or citrate buffers as well as various phosphate systems are relatively insensitive to changes in temperature. This point has to be remembered when buffers are used for the practical purpose of keeping a solution at constant pH in the laboratory and in the cold room. The considerable temperature-dependence of the ionization of many groups of amino-acid side chains must be remembered for the interpretation of experiments with enzyme molecules and it can be used for the perturbation of reaction equilibria (see Section 8.1.2).

2.2.2. Effects of Structure and Environment on Dissociation Constants

The dissociation constant of an acid depends on a large number of factors. The effects of the electronic structure, the presence of other ionic charges on the molecule, the microscopic environment within the molecule, the macroscopic environment of the solvent and steric effects all add up to determine the affinity for protons of a particular group under specified conditions. These factors will all be discussed briefly. A detailed analysis useful for elucidation of the relation between structure, environment and the reactivities of acids and bases is still the subject of many research programmes and is the key problem of the chemical mechanism of enzyme catalysis.

A comparison of the pK values of carboxylic acid groups of a number of compounds is given in Table 6.

Table 6

Acid	Formula	pK
Acetic	CH_3COOH	4·75
Chloroacetic	$ClCH_2COOH$	2·87
Dichloroacetic	$Cl_2CHCOOH$	1·48
Aminoacetic	$^+H_3NCH_2COOH$	2·35
Malonic	$^-OOCCH_2COOH$	5·70

pK refers to dissociation of this proton

Pauling (1970) has recently presented a very simple review of his explanations of the reasons why $R-C{\overset{\displaystyle O}{\underset{\displaystyle OH}{}}}$ is more readily dissociated than the corresponding alcohol $R-C\!\!\begin{array}{c}\diagup H_2\\ \diagdown OH\end{array}$; the latter would have a pK at least 6 units higher. In this admirable article on structural features affecting the energy content of related compounds, Pauling points out that the larger resonance contribution to the structure

$$R-C\!\!\begin{array}{c}\diagup O\\ \diagdown O\end{array}^{-} \quad \text{as compared with} \quad R-C\!\!\begin{array}{c}\diagup O\\ \diagdown OH\end{array}$$

stabilizes the ionized species. This decrease in energy on ionization favours the removal of the proton. Such resonance stabilization of the ion does not occur in comparable alcohols.

The substitution of hydrogen by halogen atoms results in the withdrawal of electrons towards the halo alkyl group and corresponding weakening of the H^+ binding to the carboxyl group. Changes of pKs in the direction of stronger acids (better proton donors) are observed in many reactions used for protein modification involving halide substituted compounds. The phenolic hydroxyl of tyrosine changes its pK from 10·95 to 6·48 in 2,6 diiodotyrosine. The pKs of the NH_3^+ and COOH groups on the α-carbon of tyrosine are much less affected by this substitution.

The presence of an NH_3^+ group on aminoacetic acid (glycine) exerts repulsive forces on the proton of the $-COOH$ group and decreases the pK accordingly. Conversely, the $-COO^-$ group on malonic acid attracts the proton indicated in Table 6 and causes a considerable increase in the pK of the first group titrated on addition of strong acid to $CH_2(COO^-)_2$.

Effects of the environment are best first considered in relation to macroscopic changes due to composition of the solvent. The effect of a change in the dielectric constant will be more marked in the case of a neutral acid when direct charge interaction is the principal force. In theory all dissociation constants of systems

$$AH \rightleftharpoons A^- + H^+$$

should be sensitive to ionic strength and the addition of solvents with low dielectric constants.

The ionic strength of a solution I is defined by

$$I = \tfrac{1}{2} \sum C_i Z_i^2 \tag{2.3}$$

where M is the molarity of each ionic species and Z its charge. For a 1-1 electrolyte (NaCl for example) the ionic strength is the same as the molarity, for a 1-2 electrolyte ($CaCl_2$ for example) the ionic strength is three times the molarity. This concept takes account of charge and concentration of ions and both of these have an effect on the forces between charged groups in a solution. As an example Table 7 shows the change of pK of the second ionizing group of phosphoric acid.

Table 7. pH values for equimolar mixtures of KH_2PO_4 and Na_2HPO_4. For this mixture the ionic strength is four times the molarity of each component

Ionic strength	pK	Ionic strength	pK
0·01	7·07	0·10	6·87
0·02	7·02	0·15	6·81
0·04	6·96	0·20	6·78
0·05	6·94	0·40	6·67

As the ionic strength increases more dissociation into ions occurs The thermodynamic dissociation constant $K°$ is obtained by determining the equilibrium at different ionic strength and extrapolating the results obtained to $I = 0$.

The data for the second ionizing group of phosphoric acid given in Table 7 are perhaps of even greater practical use for making up different buffers with these commonly used components than as a universally applicable result. Factors influencing the dissociation of acids are really quite complex and some neutral acids are little more affected by ionic strength than some cationic acids.

On theoretical grounds it would be expected that the addition of low dielectric solvent would have the opposite effect from the increase in ionic strength. In a solution of low dielectric constant the increase in electrostatic forces should lead to a lower dissociation constant and higher pK. Table 8 presents a comparison of the relation between dielectric constants and pKs of carboxylic groups of acetic acid and glycine in water–dioxane and water–

Table 8

Dioxane in g weight %	Ethanol in vol. %	Dielectric constant	pK acetate	pK_1 glycine
0	0	80	4·75	2·40
20	0	60	5·29	2·70
45	0	40	6·30	3·20
0	90	30	7·10	3·65
70	0	18	8·31	4·05

ethanol mixtures. The change in pK of the carboxylic group of acetic acid is in the direction and of the magnitude expected and the changes are consistent with changes in dielectric constant of the two different solvents. The smaller change in pK_1 of glycine with decreasing dielectric constant can be ascribed to the opposing action of the ammonium group on the molecule. The effect of the charge interaction with NH_3^+ will increase with decreasing dielectric constant.

The effects of changes in dielectric constants on pKs of cationic acids are generally expected to be smaller and in the direction of decreasing pK with decreasing dielectric constant. Triethanolamine and ethylamine decrease their pKs by approximately 0·6 and 0·8, respectively, in 80 per cent. ethanol as compared with aqueous solutions. However, the pK_2 of glycine, corresponding to the titration of the amino group, changes only from 9·81 in water to 9·99 in 90 per cent. ethanol while it decreases to 8·35 in 45 g weight per cent. dioxane.

Findlay, Mathias and Rabin (1962) have carried out a classical investigation on the charge types of groups on the catalytic site of pancreatic ribonuclease. The state of ionization of the groups on the enzyme in different solvent mixtures was deduced from rate measurements in buffered solutions. In this investigation as well as in the studies of Bates and Schwarzenbach (1955) the complexities of pH measurement and interpretation of results on dielectric effects in multi-component systems are well documented.

In addition to the difficulties of interpretation referred to above, changes in dielectric constant of the medium can be shielded by the microscopic environment of a group within the macromolecule. The effect of another charge nearby has been discussed in relation to the pK_1 of glycine. Furthermore, it is found that some important ionizing groups involved in catalytic mechanisms are within a non-polar environment at the active site of a protein molecule. In such a situation the addition of organic solvent will have no effect on the pK of a neutral acid. All methods for the identification of a charged group within a complex system need cautious interpretation. How results of a number of approaches to such identifications are used to give circumstantial evidence is discussed in Section 2.2.4. The theoretical background to this problem is well treated in Edsall and Wyman (1959).

2.2.3. Properties of Acidic and Basic Amino Acid Residues

The following groups with acid/base properties are found in proteins or peptides:

$$-COOH \rightleftharpoons -COO^- + H^+ \quad \begin{cases} \text{α-carboxyl end-groups} \\ \text{β-carboxyl of aspartate} \\ \text{γ-carboxyl of glutamate} \end{cases}$$

$$-NH_3^+ \rightleftharpoons -NH_2 + H^+$$

$\begin{cases} \alpha\text{-amino end-groups} \\ \varepsilon\text{-amino group of lysine} \end{cases}$

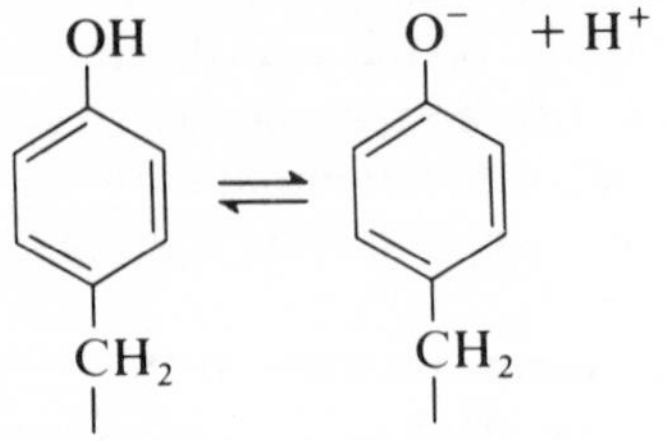

Phenolic OH tyrosine

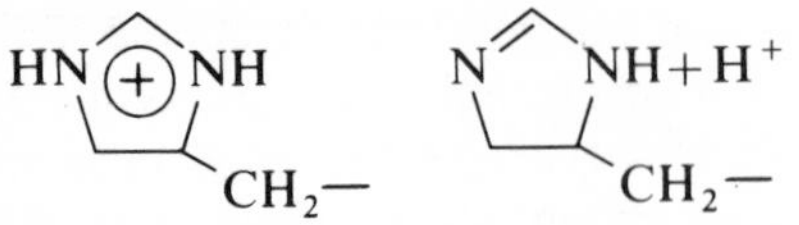

Imidazole of histidine

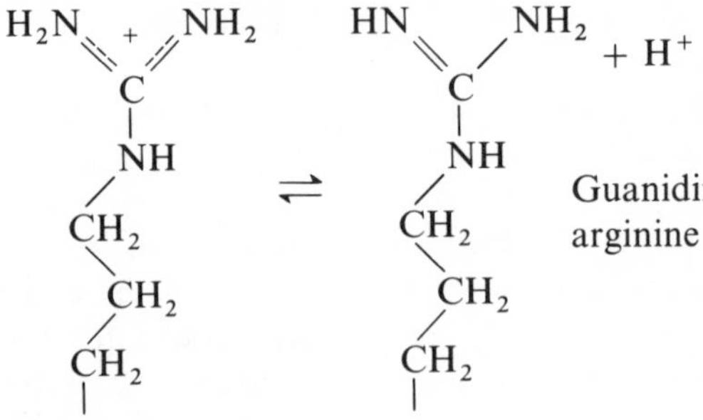

Guanidine group of arginine

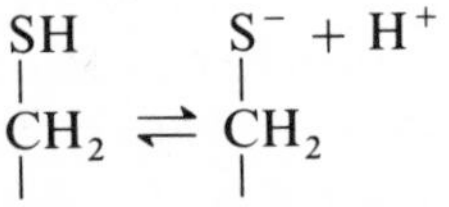

Sulphydryl group of cysteine

The above-listed groups would be expected to be the only ones to undergo changes in protonation when isolated amino acids in solutions between pH 2 to 12 are considered. However, within the special environments of catalytic sites the hydroxyl group of serine could be ionized. Tryptophan and methionine, too, may have special reactivities but need not be considered in a discussion of amino acid residues as general acids or general bases.

A solution of the amino acid alanine in pure water has a pH of 6·0. Titration of this solution with a strong acid shows the presence of one proton accepting group with pK = 2·35. Titration of another sample of the alanine solution from pH 6·0 with a strong base shows the presence of proton donor with pK 9·7.

The structure of alanine at pH 6 can be established by a number of additional experiments. In acid solution alanine will migrate towards the negative electrode of an electrophoresis apparatus while in alkaline solution it will travel towards the positive electrode of an electrophoresis apparatus. At pH 6 alanine will not move at all in an electric field; at this pH the molecule is neutral and is said to be at its isoelectric point. The zero net charge at pH 6 is due to the zwitterionic structure B:

$$
\underset{\substack{|\\ NH_3^+}}{CH_3-CH-COOH} \underset{\substack{pK\ 2.35}}{\overset{\pm H^+}{\rightleftharpoons}} \underset{\substack{|\\ NH_3^+}}{CH_3-CH-COO^-} \underset{\substack{pK\ 9.7}}{\overset{\pm H^+}{\rightleftharpoons}} \underset{\substack{|\\ NH_2}}{CH_3-CH-COO^-}
$$

A	B	C
Net charge +ve	Net charge zero zwitterion	Net charge −ve

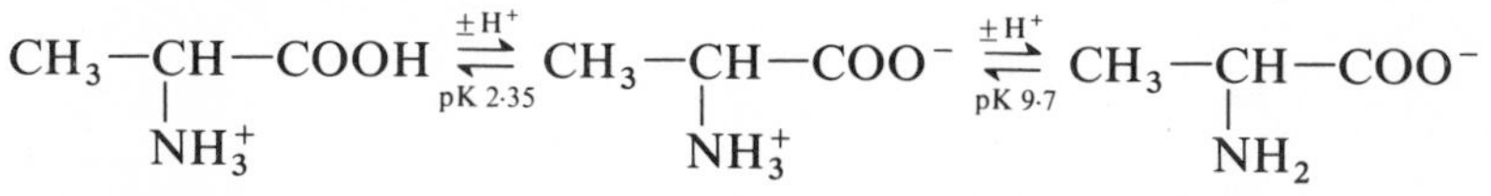

The isoelectric point is dependent on ionic strength. The isoionic point is the isoelectric point at zero ionic strength.

If the two ionizing groups at the α-carbon of an amino acid with a non-polar side chain are considered in terms of

$$
K_1 = \frac{C_{H^+}C_{RNH_3^+COO^-}}{C_{RNH_3^+COOH}}
$$

$$
K_2 = \frac{C_{H^+}C_{RNH_2COO^-}}{C_{RNH_3^+COO^-}}
$$

and $C_{H^+}^I$ is the isoelectric hydrogen ion concentration when

$$
\frac{C_{H^+}^I C_{RNH_3^+COO^-}}{K_1} = \frac{K_2 C_{RNH_3^+COO^-}}{C_{H^+}^I}
$$

then

$$
(C_{H^+}^I)^2 = K_1 K_2
$$

and

$$
2 \log C_{H^+}^I = \log K_1 + \log K_2
$$

$$
\text{pH isoelectric} = \frac{pK_1 + pK_2}{2} = pI
$$

for the three groups of glutamate or aspartate:

$$
(C_{H^+}^I)^3 = K_1 K_2 C_{H^+}^I + 2K_1 K_2 K_3
$$

where K_1 and K_2 are the dissociation constants of the two carboxyl groups and K_3 is the dissociation constant of the α-amino group.

An important property of amino acids, proteins and other molecules involved in enzyme reactions is that they are dipolar. This is defined as follows:

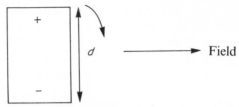

If two opposite charges on a molecule are separated by distance d and the molecule is in an electric field, there will be a torque on the molecule towards orientation in the field. If a large molecule has a large number of charges randomly distributed over its surface, resulting in zero net charge and absence of a permanent dipole, a dipolar structure can be induced by another similar molecule. Even if it is assumed that on a time average both molecules are neutral and non-polar, at any one moment in time local charges will be non-random. If a temporarily positively charged area of a molecule approaches a temporarily negative area of another molecule, the two molecules will attract each other and will favour the non-random distribution of charge in the contact area. This is similar to van der Waals' attraction of uncharged molecules, only in this case it is the distribution of electrons which is changed for the formation of an induced dipole.

The dipole moment is defined as

$$\mu = dq$$

with q the magnitude of either charge and d the distance between them. For two opposite charges on a molecule at a distance of 1 A

$$\mu = 4.8 \times 10^{-10} \text{ e.s.u.} \times 1 \times 10^{-8} \text{ cm} = 4.8 \times 10^{-18} \text{ e.s.u. cm}$$

For a water molecule $\mu = 1.9 \times 10^{-18}$ e.s.u. cm and for amino acids

$$3 \times 10^{-8} \text{ cm}$$

$\mu = 14.4$ e.s.u. cm. The dipole moment of molecules can be calculated from measurements of the dielectric constant (see Edsall and Wyman, 1959).

From the increase in the dielectric constant of water on addition of glycine a dipole moment of 20 e.s.u. cm can be deduced. This is one of many different pieces of evidence for the zwitterionic structure of amino acids in aqueous solutions.

As discussed for electrostatic interaction (Section 1.1.2) attraction between dipolar molecules are dependent on the environment. The crystal lattice of glycine is quite strong (melting point 225° C) compared with glycolic acid (melting point 63° C). In aqueous solution the attractive forces become weak and in the presence of neutral salts they are virtually abolished. Similarly, in completely salt-free solutions of isoionic serum albumin there is strong dipolar attraction between the protein molecules. On addition of sodium chloride this attractive force becomes much weaker.

2.2.4. Systems of Ionizing Groups and their Reactivities

Another important problem in the study of ionic properties of amino acids, which provides a lesson for investigations of ionic groups at the active site of enzymes, is the characterization of two neighbouring groups with pKs fairly close together. The catalytic site usually contains systems of ionizing groups rather than one isolated one. The picture of groups at the active site of a hydrolytic enzyme presented in Figure 4 illustrates this point. Much effort has been concentrated in the past on attempts to identify a particular amino acid residue from determinations of a pK for a group involved in catalysis. The perturbations of pKs of groups within complex molecules has already been referred to. The imidazole group at the catalytic site of trypsin can have pK values from 5·5 to 7·2, depending on the modification of the protein. Other dangers in assigning pK values from the pH dependence of rate constants are discussed in Section 6.5.2. In this section the assignment of pKs to groups like the $-NH_3^+$ and $-OH$ of tyrosine and the $-NH_3^+$ and $-SH$ of cysteine will be considered.

For the purpose of this discussion the carboxylic group of cysteine is ignored and four forms of the amino acid found in the pH region 7 to 12 are written as:

$$COO^-$$
$$|$$
$$HSCH_2CHNH_3^+$$
$$I$$

K_{II} K_I

$$COO^-$$
$$|$$
$$-SCH_2CHNH_3^+ \quad II$$

$$IV \quad HSCH_2CHNH_2$$
$$| \quad\quad\quad\quad\quad COO^-$$

$$III$$

K_{III} K_{IV}

$$COO^-$$
$$|$$
$$-SCH_2CHNH_2$$

The titration curve of a dibasic acid

$$AH_2 \overset{K_{1}}{\rightleftharpoons} AH^- \overset{K_{2}}{\rightleftharpoons} A^=$$

can only be interpreted in terms of two pKs if these are very far apart. If the basic form with $-S^-$ is divided among two species II and III, both at

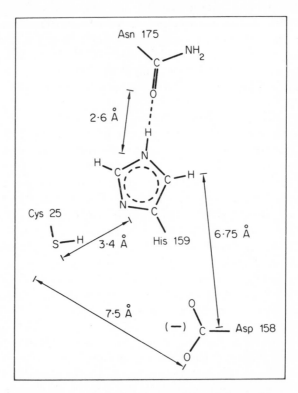

Figure 4 A schematic representation of the position of ionizing groups at the active site of papain (after Drenth et al., *Phil. Trans. Roy. Soc. Lond.*, **B257**, 231 and reproduced with permission

significant concentrations and the basic form $-NH_2$ is divided among two species III and IV, both at significant concentrations, then the ionizations of two groups are closely coupled. For such a case titration with alkali of species I results in the first equivalent of protons dissociating to form a mixture of species II and IV

$$K_1 = K_{II} + K_I$$

and the constant for the second proton

$$\frac{1}{K_2} = \frac{1}{K_{III}} + \frac{1}{K_{IV}}$$

also

$$K_1 K_2 = K_I K_{IV} = K_{II} K_{III}$$

If $K_I \gg K_{II}$ then $K_1 = K_I$

If $K_{III} \gg K_{IV}$ then $K_2 = K_{IV}$

The reason why the literature is full of attempts and criticisms of attempts to assign individual pK values to closely coupled groups lies in the experimental difficulties of studying the state of one of them without disturbing the other. This is true both for isolated cysteine or tyrosine and a site on a protein with a pair of residues of the amino acids cysteine, tyrosine, lysine, histidine or an α-amino end-group. In theory it should be possible to use rapid reaction techniques to determine the state of ionization of one group without disturbing the other significantly. Proton magnetic resonance is likely to be the method of choice for the determination of intrinsic pKs of individual groups within a system.

The state of ionization of —SH and of the phenolic —OH of tyrosine can be determined spectrophotometrically. The absorbancy of —S⁻ at 235 mμ ($\varepsilon_M = 5,000$) disappears on protonation to form —SH. The absorbancy of tyrosine —OH has a maximum at 274·5 mμ ($\varepsilon_M = 2,330$). Unfortunately this is not an absolute measure of the ionization of these residues. Spectra are perturbed by the presence of charged groups in the neighbourhood and so a change in ionization of another group can change the tyrosine spectrum. Until kinetic procedures are perfected to determine the state of ionization of individual groups it appears most sensible to talk about pKs of a system of groups as a whole.

The calculation of the relative concentrations of different species in a system

$$\begin{array}{ccc}
& \text{HAH} & \\
{}^{K_I}\diagup & & \diagdown{}^{K_{II}} \\
{}^-\text{AH} & & \text{HA}^- \\
{}_{K_{III}}\diagdown & & \diagup{}_{K_{IV}} \\
& {}^-\text{A}^- &
\end{array}$$

requires another consideration, which is best illustrated in terms of two intrinsically identical proton binding sites. If K_1 and K_2 are the two dissociation constants for the first and second proton and K_I, K_{II}, K_{III} and K_{IV}

are the intrinsic dissociation constants, clearly it is impossible to distinguish experimentally whether the left-hand or the right-hand proton of HAH dissociates. The two intrinsic constants concern the first and the second proton to dissociate, whichever this happens to be. The two macroscopic and the four intrinsic constants are defined as follows:

$$K_I = \frac{C_{-AH} C_{H^+}}{C_{HAH}} \qquad\qquad K_{II} = \frac{C_{HA^-} C_{H^+}}{C_{HAH}}$$

$$K_{III} = \frac{C_{-A^-} C_{H^+}}{C_{-AH}} \qquad\qquad K_{IV} = \frac{C_{-A^-} C_{H^+}}{C_{HA^-}}$$

$$K_1 = \frac{C_{H^+}[C_{-AH} + C_{HA^-}]}{C_{HAH}} \qquad\qquad K_2 = \frac{C_{H^+} C_{-A^-}}{C_{HA^-} + C_{-AH}}$$

For the present simplified case it has been defined that

$$K_I = K_{II} = K_{III} = K_{IV} = K_0$$

the identical intrinsic constants. It follows that

$$K_1 = K_I + K_{II} = 2K_0$$

and

$$\frac{1}{K_2} = \frac{1}{K_{III}} + \frac{1}{K_{IV}} = \frac{2}{K_0} \qquad K_2 = \tfrac{1}{2}K_0$$

and

$$K_1 = 4K_2$$

The general equation for multiple binding is derived from an extension of the statistical argument that the first proton can dissociate from one of two places but can return only to one, while the second proton can dissociate only from one site while it can return to two sites. The probability for occupying a site is four times greater for the second proton and hence its dissociation constant 1/4 compared with the first. In a polymer with n sites and equal intrinsic dissociation constants K_0, K_i for the ith proton is

$$K_i = \frac{K_0(n - i + 1)}{i}$$

The titration of two identical sites with no interaction yields one single dissociation constant equivalent to the titration of the same concentration of sites on separate molecules. The statistical correction is required to calculate the concentrations of the different species C_{HAH}, C_{-A^-} and $[C_{-AH} + C_{HA^-}]$ since the last two cannot be distinguished from each other. This type of calculation is much more important when multiple binding sites of substrates are discussed (see Section 4.1.3).

Multiple binding sites for protons are unlikely to be truly independent due to electrostatic factors discussed above. For oxalic acid $K_1/K_2 = 920$ and for malonic acid $K_1/K_2 = 650$ and even with seven carbons between the two carboxylate groups K_1/K_2 is still 6.

While the determination of pKs of groups at the active sites of proteins is not a reliable guide to their identification, it provides invaluable information about their absolute reactivity and its pH dependence. A detailed discussion of Brönsted's and Hammett's relations between dissociation constants and such properties as acid-base catalysis, nucleophilic catalysis, replacement and leaving facilities, is beyond the scope of this volume. The classical text of Hammett (1940) and the more recent discussions with examples relevant to enzyme catalysis (Bender, 1972, Jencks, 1969 and Bruice and Benkovic, 1966) should be consulted.

The Brönsted catalysis law

$$\log k = c + \beta \, pK$$

was formulated to relate the rate constant k of a reaction catalysed by an acid or base to the dissociation constant. The important constant used to distinguish between general base catalysis and nucleophilic catalysis is β. Perhaps it might be better to say that β provides a measure of the balance between nucleophilic and general base character of the catalysis. When $\beta = 1$ the process is classified as nucleophilic and covalent intermediates between the catalytic group and substrate are likely as, for instance, in the decarboxylation of acetoacetate (Westheimer, 1963):

$$
\begin{array}{ccc}
 & H & CH_3 \\
 & | & | \\
\text{Enzyme–Lysine } \varepsilon- & N\cdots C=O \\
 & | & | \\
 & H & CH_2 \\
 & & | \\
-H_2O & & COOH
\end{array}
$$

$$
\begin{array}{cc}
 & CH_3 \\
 & | \\
\text{Enzyme–Lysine } \varepsilon- \overset{+}{N}=C \\
 & | \quad | \\
 & H \quad CH_2 \\
 & \quad | \\
 & \quad COO^-
\end{array}
$$

This compound decomposes in water to acetone, CO_2 and free enzyme

When β is between 0·2 and 0·5 general base catalysis is implicated as in the imidazole catalysed transfer of acyl groups from substrates to the serine hydroxyl of proteolytic enzymes (Bender and Kezdy, 1965). While in model

reactions pKs of catalysts can be changed easily, in the enzymic examples
quoted the evidence for nucleophilic catalysis comes from the isolation of
intermediates and the absence of solvent deuterium isotope effects (see
Section 7.4.2). Recently, however, Katchalski (1970) has shown how the
micro environment of catalytic groups, and with that their pKs, can be
changed.

The Brönsted relation between pK and nucleophilic catalysis has an
important bearing on the effectiveness of a group as a catalyst at a particular
pH. If two groups, which make cationic acids with pK values of 6 and 9,
respectively, are compared at pH 6, they will be nearly equally effective.
The low concentration of the basic form of the group with pK 9 is compensated for by its greater catalytic power.

2.2.5. Properties and Uses of Indicators

Many phenolic dyes are weak acids/bases, which have very large difference
spectra between their two ionized forms. They have for a long time been used
for simple, rapid and approximate pH determinations. Modern spectro-
photometric equipment has made these dyes useful for quite sophisticated
experiments (Darrow and Colowick, 1962; Chance, 1969).

Table 9

pK	Compound	MW	$m\mu$	ε_1 mole cm
1·4	Thymol blue	466	550	33×10^3
2·6	Benzyl orange	405	505	56
4·0	Bromo Cl phenol blue	569	590	72
4·7	Bromo cresol green	698	610	42
5·9	Chloro phenol red	423	580	44
6·9	Bromthymol blue	624	620	36
7·0	p-nitrophenol	139	400	18·3
7·9	phenol red	354	560	54
8·1	cresol red	382	525	53
9·0	Thymol blue	466	600	30
9·6	Phenolphthalein	318	552	33

The data presented in Table 9 are intended to be a guide to the use of
different indicators for monitoring changes in the concentration of hydrogen
ions. The basic forms of the indicators absorb light at the wavelengths
listed. Precise data have to be obtained from calibrations of the particular
indicator used under the conditions of the experiment. Table 10 gives the
relative concentrations of the acidic and basic forms of a compound as a
function of the difference pK–pH. These values are useful for the calculation

Table 10

pK–pH	A	B
−4	99·99	0·01
−3	99·94	0·06
−2	99·02	0·98
−1·5	96·95	3·05
−1·2	94·06	5·94
−1·0	90·91	9·09
−0·9	88·82	11·18
−0·8	86·30	13·70
−0·7	83·36	16·64
−0·6	79·96	20·04
−0·5	75·99	24·01
−0·4	71·50	28·50
−0·3	66·61	33·39
−0·2	61·30	38·70
−0·1	55·72	44·28
0	50·0	50·0
0·1	44·28	55·72
0·2	38·70	61·30
0·3	33·39	66·61
0·4	28·50	71·50
0·5	24·01	75·99
0·6	20·04	79·96
0·7	16·64	83·36
0·8	13·70	86·30
0·9	11·18	88·82
1·0	9·09	90·91
1·2	5·94	94·06
1·5	3·05	96·95
2	0·98	99·02
3	0·06	99·94
4	0·01	99·99

of changes in absorbancy to be expected with changes in pH when different indicators are used. The table should also prove useful for the preparation of buffers at a prescribed pH from weak acids or bases with known dissociation constants.

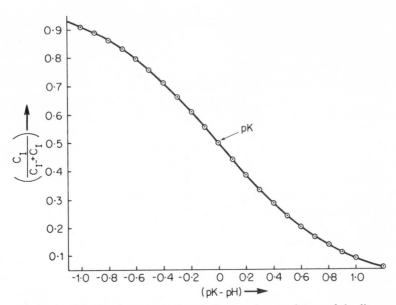

Figure 5   This titration curve of an indicator gives a picture of the linear region for change in the concentration C_I (usually the coloured form of the indicator) over the total indicator concentration

Inspection of Figure 5 shows that a change of 0·1 pH unit in the region of pK $\pm$ 0·3 causes a 5·5 per cent. change in the proportion of the basic (coloured) form of an indicator. This makes indicators very suitable for monitoring small pH changes as a function of time in a recording spectro-photometer. A kinetic experiment with a total change of 0·1 pH can be recorded accurately. The absolute sensitivity to proton uptake or liberation by the reactants does, of course, depend on the pH of the experiment. In the region pH 6·5 to 8·5, where most enzyme reactions are studied, CO_2 has to be carefully excluded from contact with the solutions (see Section 2.2.6).

Among the points to be considered when indicators are used to follow reactions in complex systems are the rates of the indicator reaction itself and the influence of the solution on the properties of the indicator. The relaxation times of protonation equilibria are very short (Eigen, 1963) and need not be taken into account for most reactions. In the case of 10 μM phenol red at pH 8 the time constant for approach to a new equilibrium, after an in-stantaneous change in C_{H^+}, is approximately 10 μsec.

The effects of other components of the solution (ionic strength, dielectric constants) on the pKs and spectra of indicators have to be determined in control experiments. The presence of proteins can have some special effects which are very interesting in some cases and make the method unusable in others. It seems likely that in solutions containing indicator and protein there will be some binding of the indicator to the protein. This phenomenon can cause serious artifacts. If the spectrum and the pK of the indicator is unchanged by the presence of the protein and the reaction studied is not affected by changes in the indicator concentration, then this artifact is not likely to be serious.

The binding of indicators and other dyes to specific sites on a protein can, however, provide interesting information. For instance, Antonini et al. (1963) intended to use bromthymol blue to study the kinetics of the well known 'Bohr effect' (see p. 32), the release of 2·7 moles of protons during oxygenation of 1 mole of tetrameric haemoglobin at physiological pH. However, they found that a haemoglobin molecule bound at least 10 molecules of bromthymol blue per mole of haem and that the pK of protein bound indicator was about 1·0 unit higher than that of free indicator. Similar changes in the properties of bromthymol blue are found in the presence of serum albumin. Adsorption of the indicator to non-polar areas on the protein surface can account for the change in pK.

An interesting feature of the adsorption of bromthymol blue on haemoglobin is that the binding energy of the dye depends on the state of the protein; oxygenation and the consequent conformation change reduce the binding constant. While it is beyond the scope of this section to discuss the detail of the reactions of haemoglobin, it should be pointed out that indicators which do bind to proteins can provide information about structural changes.

Phenol red was subsequently used to study the kinetics of the liberation and uptake of the 'Bohr' protons. This indicator has proved suitable for kinetic investigations of many reactions which result in the change of the hydrogen ion concentration due to the stoichiometry of the substrate-product reaction or due to changes in ionization of the protein during enzyme-substrate complex formation.

2.2.6. The Formation and Properties of Carbonic Acid

CO_2, like protons, is a substrate or product of many enzyme reactions. The rates and equilibria of the consecutive processes occurring near neutral pH

$$CO_2 + H_2O \rightleftharpoons H_2CO_3 \qquad K_H = 2 \cdot 58 \times 10^{-3}$$

$$H_2CO_3 \rightleftharpoons HCO_3^- + H^+ \qquad K_A = 2 \cdot 5 \times 10^{-4}$$

are, therefore, of considerable importance in the study of mechanisms.

The normal atmosphere contains 0·03 per cent. (by volume) CO_2. The partial pressure of CO_2 under normal aerobic conditions is 3×10^{-4} atm. The solubility of CO_2 at 25° in pure water (it is decreased by the presence of salts) is 0·0344 M in the presence of 1 atm. of CO_2 and 10·35 μM in the presence of the partial pressure of CO_2 under aerobic conditions. The second acid dissociation constant

$$HCO_3^- \rightleftharpoons CO_3^= + H^+ \qquad\qquad pK = 10\cdot33$$

is omitted from consideration in the present discussion. In Table 11 the solubilities of some gases are given as functions of their partial pressure.

Table 11. Solubility in Pure Water of Gases at 25°

At 25° and 1 atm. of pressure 1 mole of gas has a volume of 24·45 l.

Gas	N_2	O_2	CO_2	CO
Volume % in air	78·09	20·95	0·03	—
Partial pressure at 1 atm. of air	0·78 atm. 594 mm Hg	0·21 159·2	$0\cdot3 \times 10^{-3}$ 0·228	— —
Solubility under 1 atm. of air pressure	$1\cdot11 \ 10^{-2}$ l./l. $4\cdot55 \times 10^{-4}$ M	$5\cdot9 \times 10^{-3}$ $2\cdot4 \times 10^{-4}$	$2\cdot53 \times 10^{-4}$ $1\cdot035 \times 10^{-5}$	— —
Solubility per mm Hg pure gas	$1\cdot88 \times 10^{-5}$ l./l. $0\cdot77 \times 10^{-6}$ M	$3\cdot72 \times 10^{-5}$ $1\cdot5 \times 10^{-6}$	$1\cdot00 \times 10^{-3}$ $4\cdot5 \times 10^{-5}$	$2\cdot82 \times 10^{-6}$ $1\cdot15 \times 10^{-6}$

The total carbon dioxide content of a solution in equilibrium with a given pressure of CO_2 and its distribution between the three species

$$C_{CO_2} + C_{H_2CO_3} + C_{HCO_3^-}$$

is determined by the pH of the solution. The apparent first pK_A calculated from the equilibrium

$$K_A = \frac{C_{H^+} \, C_{HCO_3^-}}{C_{CO_2} + C_{H_2CO_3}} = 4\cdot45 \times 10^{-7} \, M$$

is 6·35. The true $pK_a = 3\cdot60$ for the reaction

$$H_2CO_3 \rightleftharpoons HCO_3^- + H^+$$

can be calculated from the apparent constant and $K_H = 2\cdot58 \times 10^{-3}$ (at 25° with pure water being at unit concentration, see Section 3.1.1) for the hydration of CO_2.

The above information is of general practical use for the interpretation of the effects of ambient CO_2 on the pH of solutions and for the preparation of carbonate buffers. It now remains to discuss the reactions of CO_2 and carbonate when they are involved in enzymic reactions.

The hydration of CO_2 and the dehydration of H_2CO_3 are slow reactions compared with accompanying changes in ionization. This means that the hydration/dehydration kinetics can be followed by monitoring pH changes and that it is possible to determine whether CO_2 or HCO_3^- is the substrate for an enzyme reaction which is fast compared with the interconversion between the two forms. Both these topics will be discussed briefly.

When a slightly acid solution of CO_2 is adjusted to pH 8 the rate-limiting step for the formation of $HCO_3^- + H^+$ is the hydration process $k = 0.038$ sec^{-1}. This reaction with a half-time of $18 \, sec^{-1}$ (see Section 6.1.2) is convenient for testing methods for moderately fast pH measurements with indicators or electrodes.

The adjustment of a bicarbonate solution pH 8 to pH 6.35 results in a negligible proton uptake by HCO_3^- (pK = 3.6) prior to the rate-limiting dehydration. The pH independent first-order rate $H_2CO_3 \rightarrow CO_2 + H_2O$ is $30 \, sec^{-1}$ at 25° (see also Ho and Sturtevant 1963, and Gibbons and Edsall 1963).

The increase in capacity for transporting CO_2 and carbonate in blood after oxygen has been removed from haemoglobin is due to the uptake of protons by deoxyhaemoglobin from the solution and to the chemical combination between amino end-groups of the protein and CO_2 to form carbamates:

$$R - NH_2 + CO_2 \rightleftharpoons R - NHCOO^- + H^+$$

the pKs of carbamates are approximately 5.2.

This reaction is discussed in some detail by Roughton (1970) and Kilmartin, and Rossi–Bernardi (1969, 1971).

Carbonic anhydrase is an enzyme which occurs in red blood cells together with haemoglobin, and which catalyses the hydration–dehydration reactions. With a turnover number (see p. 128) of about 10^6 moles of CO_2 hydrated per mole of enzyme per second it is one of the fastest enzymes known. This is an essential part of the control of O_2 versus CO_2 exchange in the bloodstream. The enzyme also turns out to be very useful as an analytical tool. A few reactions catalysed by different enzymes have been studied in the presence and absence of carbonic anhydrase:

(1) $\underset{NH_2 \quad NH_2}{\overset{\overset{\displaystyle O}{\overset{\|}{C}}}{\diagup \diagdown}}$ $+ \, H_2O \rightleftharpoons CO_2 + 2\,NH_3$ Urease
(Krebs and Roughton, 1948)

(2) $\begin{array}{c} CH_3 \\ | \\ C=O \\ | \\ COOH \end{array} \rightleftharpoons \begin{array}{c} CH_3 \\ | \\ C \\ \diagup \diagdown \\ H \quad O \end{array} + CO_2$
 Yeast carboxylase
(Krebs and Roughton, 1948)

(3) $2\,CHO.COO^- + H^+ \rightleftharpoons CHOCHOHCOO^- + CO_2$ Glyoxylate
carboligase

(4) D-Ribulose 1,5 diphosphate $+ CO_2 + H_2O \rightleftharpoons$ Ribulose

2,3 diphosphoglycerate $+ 2H^+$ diphosphate-
carboxylase

(5) 6-Phosphogluconate $+ NADP^+ \rightleftharpoons$

ribulose 5-phosphate $+ NADPH + CO_2 + H^+$

(Villet and Dalziel, 1969)

Hall, Vennesland and Kezdy (1969) who studied the glyoxylate carboligase reaction in detail also summarize the results of those who have studied the other reactions. In every one of these cases the rate of formation or utilization of carbonate was dependent on the presence of carbonic anhydrase. This indicates that CO_2 is the primary product or active substrate (see, for instance, Figure 6).

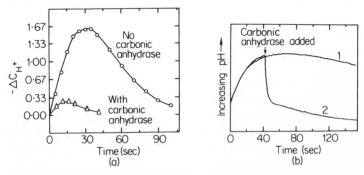

Figure 6   Hall, Vennesland and Kezdy (1969) followed the reaction

$$2\,CHOCOO^- + H^+ \rightarrow CHO.CHOCOO^- + CO_2 \qquad (1)$$
$$CO_2 + H_2O \rightarrow HCO_3^- + H^+ \qquad (2)$$

catalysed by glyoxylate carboligase *Escherichia Coli* in the presence and absence of carbonic anhydrase, which catalyses the reaction $CO_2 + H_2O \rightarrow HCO_3^- + H^+$. Graph (a) shows experiments in which proton uptake was followed spectrophotometrically in the presence of cresol red, and graph (b) shows proton uptake recorded potentiometrically. Both experiments show that free CO_2 is liberated by carboligase. In the absence of carbonic anhydrase the second reaction is slow

2.3. OXIDATION-REDUCTION PROCESSES

2.3.1. Free Energy and Oxidation-Reduction Potentials

The principles of the energy balance in oxidative reactions have to be well understood to make it possible to interpret the reactions of the many enzymes and electron transfer proteins which are at present being studied in detail.

The oxidation-reduction potential is defined as

$$E = E° - \frac{RT}{nF} \ln \frac{C_{reduced}}{C_{oxidized}}$$

in terms of the ratio of the concentrations of the oxidized and reduced forms. $E°$ is the standard oxidation-reduction potential when all reactants are at unit concentration (in their standard state; see Section 1.3.4). F is the Faraday constant (23,000 cal/V) and n is the number of electrons transferred in the process per mole equivalent. When the system is at equilibrium $E = 0$ and

$$E° = \frac{RT}{nF} \ln \frac{\bar{C}_{reduced}}{\bar{C}_{oxidized}}$$

($\bar{C}$ are equilibrium concentrations)

$$E° = \frac{0·06}{n} \log \frac{\bar{C}_{reduced}}{\bar{C}_{oxidized}} \quad \text{at } 298° \text{ K}$$

It is a convention, now almost universally accepted, that the better the electron acceptor the more positive is its potential. The potentials are usually quoted in volts and 1 electronvolt is equivalent to 23 kcal or 96·23 kjoules (see Section 1.1.1). This permits the interconversion of oxidation-reduction potential differences to units used in this and the next chapter to measure free energy changes.

In the subsequent discussion potentials will be quoted for systems at pH 7 ($E°'$) at 25°. The scale is usually fixed with reference to the standard hydrogen electrode ($E° = 0$)

$$E = \frac{RT}{F} \ln \frac{C_{H^+}}{pH^{\frac{1}{2}}}$$

with the standard pressure of hydrogen gas at unity (1 atm.)

$$E = 0·06 \log C_{H^+}$$

which gives for the potential of the hydrogen electrode at pH 7:

$$E°' = -0·42 \text{ V}$$

For the other end of the scale required to cover the range of potentials found
in enzyme reactions the ultimate electron acceptor, oxygen, gives at pH 7:

$$E^{\circ\prime} = +0\cdot82 \text{ V}$$

2.3.2. Electron Transfer Sequences

Usually the interesting process in a biological system is a coupled process
like

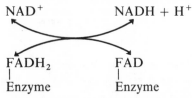

which is encountered as part of a sequence of reactions in mitochondrial
respiration. The values for $E^{\circ\prime}$ given in Tables 12 and 13 are not intended to
be taken as absolute and reliable. There are quite considerable differences in
the values quoted by different authors for the potential of the same compound.
The variation is, however, not large enough to change the position in the
ordered sequence given. Furthermore a much larger change, which can alter
this order, can be due to the inclusion of a compound into a complex structure. The simplest example is the change in potential of free flavine adenine
dinucleotide or flavine mononucleotide ($-0\cdot2$ V) to flavoprotein ($0 \pm 0\cdot2$ V).
Similarly it is interesting to compare the potential of Fe^{2+}/Fe^{3+} as the free
ion with that of Fe^{2+}/Fe^{3+} in different environments (see Table 12).

Table 12

System	$E^{0\prime}$ V at pH 7
Fe^{2+}/Fe^{3+}	$+0\cdot77$
Cyanide Fe^{2+}/Fe^{3+}	$+0\cdot36$
Haemoglobin Fe^{2+}/Fe^{3+}	$+0\cdot17$
Myoglobin Fe^{2+}/Fe^{3+}	$+0\cdot05$
Haem Fe^{2+}/Fe^{3+}	$-0\cdot12$

A list of oxidation-reduction potentials of the many different iron porphyrin
compounds involved in cytochrome systems of electron transfer would also
show a wide range of values reflecting the environmental effects on reactivity.
The study of the effects of binding FAD, FMN, NAD$^+$, Fe porphyrins,
transition metals and other cofactors to their specific sites on proteins is a
very interesting and rather neglected one. It is probably easier to get information about the chemical events in such processes than in those involving
continuous turnover of a substrate reacting at an active site on a protein.

Table 13. Oxidation-reduction potentials
at 25° and pH 7

System	$E^{0\prime}$ in volts.
Hydrogen	-0.42
NADH/NAD	-0.32
Flavine nucleotide (free)	-0.20
Acetaldehyde/ethanol	-0.20
Pyruvate/lactate	-0.18
Oxaloacetate/malate	-0.17
Flavoproteins	~ 0.00
Cytochrome b	~ 0.00
Cytochrome c	$+0.26$
Cytochrome a	$+0.28$
Oxygen	$+0.82$

The flow of electrons in the oxidation of substrates like malate, lactate or ethanol is illustrated by the following scheme:

$$\text{Substrate} \rightarrow \text{NAD}^+ \rightarrow \text{Flavoprotein} \rightarrow \text{cyt c} \rightarrow \text{cyt a} \rightarrow \tfrac{1}{2}\text{O}_2$$

$$\sim -0.18 \qquad -0.32 \qquad 0 \qquad +0.26 \qquad +0.28 \qquad +0.82$$

This scheme is oversimplified in several respects; other carriers occur in between those quoted and the stoichiometry must change from 1 to 2 moles of carrier either at the flavoprotein level or just after it.

It is, however, clear from the above scheme that the electron flow from NAD^+ to O_2 is along a path of increasing electron affinity. This is not the case during the transfer from substrates to NAD^+. This deserves some further discussion. First, this example can be used for a numerical illustration. The difference of 0.15 V between the oxaloacetate/malate and the NADH/NAD$^+$ couple at pH 7 corresponds to a standard free energy difference of 3.45 kcal and from

$$\Delta G^\circ = -RT \ln K \qquad \text{(see Section 1.3.4)}$$

$$\log K = 2.52$$

$$K = \frac{C_{\text{oxaloacetate}} C_{\text{NADH}}}{C_{\text{malate}} C_{\text{NAD}^+}} = 715 \text{ (at } C_{\text{H}^+} = 10^{-7})$$

Again it is not necessary to assume that the absolute values quoted are reliable to see that the product of the concentrations of malate and NAD$^+$ must be nearly a thousand times that of the product of the concentrations of oxaloacetate and NADH for the oxidation of malate to proceed.

In a structurally and kinetically organized system like mitochondria several factors favour the oxidation of substrates. The concentration of free NADH is maintained very low first by its tight binding to the high concentration of binding sites and secondly by the rapid oxidation of any free NADH produced. Both these factors will be discussed in the next chapter.

Some of the NADP-linked substrates have potentials which favour nucleotide reduction on equilibrium considerations alone:

$$\text{isocitrate} \qquad\qquad E^{\circ\prime} = -0\text{·}375$$

$$\text{6-phosphogluconate} \qquad E^{\circ\prime} = -0\text{·}346$$

In both these reactions the oxidation is coupled to decarboxylation. The oxidation of glucose 6-phosphate to the lactone ($E^{\circ\prime} = -0\text{·}280$) is favoured by rapid product removal through the hydrolysis of the lactone to 6-phosphogluconate.

For oxidation-reduction processes which do not involve the liberation or uptake of a proton

$$\Delta G^{\circ} = \Delta G^{\circ\prime} \qquad \text{and} \qquad E^{\circ\prime} = -\Delta G^{\circ}/nF$$

The following calculation should illustrate the principle of comparing $E^{\circ\prime}$ values with the standard free energy ΔG° and the standard free energy at pH 7 $\Delta G^{\circ\prime}$. For a two-electron transfer process such as is involved in the oxidation of $\text{NADH} + \text{H}^{+} \rightarrow \text{NAD}^{+} + \text{H}_2$ gas:

$$E^{\circ\prime} = -(\Delta G^{\circ} + RT \ln 10^{7})/2F$$

and

$$\Delta G^{\circ} = \Delta G^{\circ\prime} - RT \ln 10^{7}$$

CHAPTER 3

Chemical Equilibria

3.1. HYDROLYSIS AND TRANSFER

3.1.1. Standard States

The free energy change of a chemical reaction depends on the standard free energy of the reaction and on the absolute concentrations of the reactants. The standard free energy change $\Delta G°$ was defined in Section 1.3.4 as the change when 1 mole reactant is converted into 1 mole of product when both reactants and products are maintained in their standard state. Normally the standard state is chosen as a 1 molar solution or a solution of unit activity (see Section 1.3.3). For some substances other special reference states are chosen. For instance, for water, which is 55·5 M for dilute aqueous solution, the standard state is often taken as pure water. As long as the condition is defined that pure water is taken as unit activity, this results in an unambiguous table of data for comparative purposes. For instance, this convention is used when the free energies of hydrolysis of a series of compounds are compared.

There are comparisons of free energies for which it is more realistic to take 1 M water as the reference state. The difference between writing an equilibrium constant for the same reaction with the two standard states is $\ln(1/55·5)$, which corresponds for the case of hydrolysis reactions to an additional 2·38 kcal.

Another component for which a special standard state is at times defined is the hydrogen ion. For a hydrolysis reaction of the type

$$R_1COOR_2 + H_2O \rightleftharpoons R_1CO_2^- + R_2OH + H^+$$

the stoichiometry of H^+ depends on the dissociation constant

$$K = \frac{C_{R_1CO_2^-} C_{H^+}}{C_{R_1CO_2H}}$$

and on the pH. The pH dependence of ester hydrolysis as well as the more complex case when a neutral reactant is converted into

$$R_1CO_2^- + R_2NH_3^+$$

can be described by

$$K_{app} = K_{hydr}(1 + C_{H^+}/K_1)(1 + K_2/C_{H^+})$$

where K_{app} is the apparent and K_{hydr} the pH independent equilibrium constant for the formation of the fully ionized products. K_1 and K_2 are the dissociation constants of the acid and base formed on hydrolysis. If the dissociation constants of two carboxylic acids are K_A^A and K_A^B the equilibrium

$$A \rightleftharpoons B$$

of interconversion is dependent on pH in the following way

$$\frac{C_B^T}{C_A^T} = \frac{C_{B^-}}{C_{A^-}} \frac{K_A^A(C_{H^+} + K_A^B)}{K_A^B(C_{H^+} + K_A^A)}$$

Several other cases of the effects of pH on equilibria are discussed by Krebs (1953). The principle involved in all these derivations is simply that when substrates or products can exist in several forms, at equilibrium all forms must be in equilibrium. This point is discussed in several other sections (see, for instance, Section 3.2.3).

The special problems of the pH dependence of the hydrolysis of ATP and of the transfer of phosphate will be discussed in the next section.

The convention used in the biochemical literature has already been defined in Section 2.3.1, when pH-dependent electron transfer equilibria were discussed. The standard free energy change ΔG° is defined for unit C_{H^+} or more correctly a_{H^+}, while $\Delta G^{\circ\prime}$ is the free energy change at pH 7 with all other components at unit concentration.

The dilemma between attempting to use the correct physico-chemical definitions in terms of activity and the practical approach of using concentrations comes from a shortage of data. The measurement of pH with suitable standards results in values in terms of a_{H^+}, while all other components are usually given in 'weighed' concentrations. This is justifiable as long as one is aware of the fact that the equilibrium constants are not necessarily independent of absolute concentrations. The purist approach does not get one very far in the interpretation of biological processes since data for the precise activities in the environment of the system are difficult to come by and conditions in biological systems are usually far from ideal. This is why very precise data for the thermodynamic parameters of biochemical reactions are not so very important. The hope is that the relative values are not changed significantly in the real system.

3.1.2. The Hydrolysis of ATP and Phosphate Transfer

The thermodynamic parameters of the reaction

$$ATP^{4-} + H_2O \rightleftharpoons ADP^{3-} + HPO_4^{2-} + H^+$$

have been discussed in a large number of publications since their important position in the energy balance of biochemical reactions was highlighted by

Lipmann's review in 1941. Inspection of Table 14 shows that the standard free energy change on hydrolysis of the γ phosphate bond of ATP is roughly in the middle between the values obtained for hexose monophosphates and phosphoenol pyruvate or 1,3-diphosphoglycerate. ATP is therefore in a position to act as a buffer for the distribution of phosphoryl groups among the essential phosphorylated intermediates. The local concentration of hydrogen and divalent cations will have a marked effect on the observed equilibria of all these reactions since the dissociation constants of the reactants on the two sides of the equations are different. The largest effects are found for the equilibrium between ATP and ADP plus orthophosphate.

Table 14. Standard free energy changes at pH 7 $\Delta G^{\circ\prime}$ at 25° for some reactions selected for the discussion of enzyme catalysed processes

Hydrolysis of phosphate esters (Pi stands for orthophosphate)	$\Delta G^{\circ\prime}$ kcal
1,3-diphosphoglycerate $\rightarrow$ 3-phosphoglycerate + Pi	-13.6
Phosphoenol pyruvate $\rightarrow$ pyruvate + Pi	-13.3
Creatine phosphate $\rightarrow$ creatine + Pi	-10.2
Acetyl phosphate $\rightarrow$ acetate + Pi	-10.1
Adenosine triphosphate $\rightarrow$ adenosine phosphate + pyrophosphate	-8.0
Adenosine triphosphate $\rightarrow$ adenosine diphosphate + Pi	-7.7
Adenosine diphosphate $\rightarrow$ adenosine phosphate + Pi	-6.6
Pyrophosphate $\rightarrow$ 2 Pi	-6.6
Glucose 1-phosphate $\rightarrow$ glucose + Pi	-5.0
Glucose 6-phosphate $\rightarrow$ glucose + Pi	-3.3
Fructose 6-phosphate $\rightarrow$ fructose + Pi	-3.1
Glycerol 1-phosphate $\rightarrow$ glycerol + Pi	-2.3
Some other reactions	
Acetyl coenzyme A $\rightarrow$ acetate + coenzyme A	-8.0
$NAD^+ + H_2$ (gas) $\rightarrow$ NADH + H^+	$+4.33$
Ethanol $\rightarrow$ acetaldehyde	$+9.68$

For these two reduction processes the free energy values should be considered together with the calculations in Section 2.3.1.

Thermodynamic data are now frequently quoted in joules (see Section 1.1.)

The standard free energy of the hydrolysis of ATP (pH 0) is in fact very small (0.3 kcal) and the relatively large free energy change at pH 7 is almost entirely due to the formation and ionization of an additional acidic group. One of the two current proposals for the mechanism of the synthesis of ATP from ADP and orthophosphate during respiration (oxidative phosphorylation) is based on the possibility that local high concentrations of hydrogen ions could occur (Mitchell, 1966).

The possible effects of changes of Mg^{2+} ion concentrations on the equilibrium constant can be illustrated by writing out dissociation constants for the reactant-Mg complexes at pH 8 (the charge balance is omitted in this scheme).

$$\text{ATP Mg} \quad \rightleftharpoons \quad \text{ADP Mg} \quad + \text{Pi Mg}$$
$$\updownarrow K_1 \qquad\qquad \updownarrow K_2 \qquad\qquad \updownarrow K_3$$
$$\text{ATP} + \text{Mg} \rightleftharpoons \text{ADP} + \text{Mg} + \text{Pi} + \text{Mg}$$

$$K_1 = 10^{-4}\,\text{M} \qquad K_2 = 7\cdot2 \times 10^{-4}\,\text{M} \qquad K_3 = 1\cdot3 \times 10^{-2}\,\text{M}$$

The combination of proton and Mg dissociation phenomena presents a formidable numerical problem. Alberty (1969) has provided a solution in terms of contour maps relating free energy to pH and pMg ($-\log C_{Mg^{2+}}$).

Unfortunately a misleading name (high-energy bond) has resulted in much misunderstanding and semantics. Lipmann himself realised that some term like transfer potential is more suitable than bond energy. Clearly, the hydrolysis of a phosphate ester involves a number of bond breaking and making steps and the free energy refers to a combination of all of these. Bond energy has a clear and quite different meaning, namely the energy required to separate two bonded atoms to infinite distance. Hydrolysis is a special case of phosphate transfer (to water as the acceptor) and results in the lowest free energy level and the free energy of hydrolysis should be taken as the transfer potential.

In the presence of the specific enzyme for the reaction phosphate is rapidly transferred between the $\text{ATP} \rightleftharpoons \text{ADP}$ couple and respective other couple; for instance, in the presence of hexokinase the following reaction occurs:

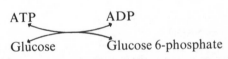

During substrate-linked (see p. 63 for the reaction catalysed by glyceraldehyde 3-phosphate dehydrogenase) as distinct from oxidative phosphorylation ATP is synthesized from ADP through phosphate transfer from phosphoenol pyruvate or 1,3-diphosphoglycerate, while the compounds of lower phosphate transfer potential (see Table 14) than ATP are phosphorylated by it. The mechanism of the supply of energy for muscle contraction is not yet understood. Considerable effort is being made to elucidate the role of ATP in this mechanism by the application of techniques developed for enzyme reactions to the study of the actomyosin system. Creatine phosphate and arginine phosphate are implicated as a phosphate store of high transfer potential in different muscles. When ADP is formed during some stage in the cycle of muscular activity, it is rapidly phosphorylated by the reserve of guanidine phosphate. The equilibrium of phosphate transfer between the adenine nucleotide and guanidine derivatives in muscle is controlled by the Mg^{2+} or Ca^{2+} concentration.

A plausible hypothesis linking enzyme mechanisms to the thermodynamic concept of transfer potential was developed (Gutfreund, 1959). This is based on the preservation of the potential during transfer via the enzyme substrate complex. This conservation of the potential can occur either through the formation of phosphoryl enzyme (EP)

$$R_1P + E \rightleftharpoons EP + R_1$$

$$EP + R_2 \rightleftharpoons E + R_2P$$

or through direct transfer of phosphate from R_1 to R_2 on the catalytic site. Either mechanism excludes the formation of inorganic phosphate and consequent loss of free energy.

The free energies of the formation or hydrolysis of the intermediates between enzymes and substrates are not yet known. Their determination depends on experiments designed to explore the structure of the intermediates; a topic on which there is as yet more speculation than real evidence. Mixed anhydride and thioesters are often involved in enzyme-substrate interaction. No doubt the special reactivity of groups within the environments of the catalytic sites also has some interesting consequences on the free energies, which are yet to be explored.

The phosphorylation of glucose by ATP (catalysed by hexokinase or glucokinase) or of fructose 6-phosphate to form the 1,6-diphosphate (catalysed by phosphofructokinase) are often referred to as irreversible reactions. From the viewpoint of metabolic regulation this is perhaps a justifiable term. In considerations of the physical basis of enzyme action it is not correct to divide processes into reversible or irreversible ones. As will be seen in Sections 3.2.1 and 3.2.2, some enzyme reactions involving much larger free energy changes can proceed in either direction in metabolic processes when conditions in the system change.

The equilibrium of the hexokinase reaction and the rates in both directions can be determined by studying the rate of exchange of P^{32}-labelled inorganic phosphate into ATP in the presence of enzyme and glucose.

3.1.3. General Considerations of Equilibria of Transfer Processes

A large proportion of the chemical processes occurring in biological systems are transfer reactions: the transfer of electrons, hydrogen, phosphoryl, acyl, carboxyl, amino groups, etc. The particular case of phosphate transfer was considered in the previous section while more general features will be illustrated in this section by discussion of some other specific examples.

The thermodynamic parameters of a reaction also provide important information for the planning of enzyme experiments. If the reaction has a negative free energy change of 3 kcal or more, so that it proceeds to more than 99 per cent. completion, then the reverse reaction need not be considered for the correction of initial rate data (see p. 131). Such is the case for the reaction catalysed by hexokinase or the hydrolytic reactions catalysed by peptidases, phosphatases, etc. It is possible to study transfer processes other than hydrolysis in the energetically unfavourable direction by starting off with very high substrate concentration and measuring the rate of appearance of very small product concentration. In the case of hydrolysis the inherently high concentration of water makes the study of the reverse reaction difficult. Lowering the concentration of water by the addition of other solvents is likely to change many parameters of the reaction. In the next section will be discussed the special features of the reversibility of reactions like that catalysed by aldolase (fructose 1,6-diphosphate $\rightarrow$ glyceraldehyde 3-phosphate + dihydroxyacetone phosphate) where there is one reactant on one side and two on the other side of the equation.

When perceptibly reversible reactions are studied it is essential to take into account not only the dependence of the rate on the substrate concentration but also the influence of the reverse reaction as the product concentration builds up. The necessary rate equations are derived in Section 6.3.5.

Practical problems connected with the study of some enzyme-catalysed reactions in both directions are discussed briefly in the rest of this section and some more examples are given in the next section.

The enzyme glutamate–oxaloacetate transaminase catalyses the reaction

$$\text{L-Glutamate} + \text{oxaloacetate} \rightleftharpoons \alpha\text{-oxoglutarate} + \text{L-aspartate}$$

the equilibrium constant for this reaction at neutral pH and 25° is K = 6·74. This process can be studied in both directions and the partial reaction of either glutamate or aspartate with the enzyme can also be followed (Fasella, 1967). The enzyme consists of a dimeric protein with two identical subunits and each subunit has a pyridoxal phosphate group attached to it. The partial reaction proceeds as follows

$$\text{Enzyme-pyridoxal} + \text{glutamate} \rightleftharpoons \text{enzyme-pyridoxamine} + \alpha\text{-oxoglutarate}$$

Enzyme-pyridoxamine can react with either α-oxoglutarate or oxaloacetate to form enzyme-pyridoxal and the respective α-amino acid.

Another example which illustrates some important points is the reaction catalysed by lactate dehydrogenase:

$$\text{Lactate} + \text{NAD}^+ \rightleftharpoons \text{pyruvate} + \text{NADH} + \text{H}^+ \text{ at } 25° \text{ K} = 3\cdot7 \times 10^{-12} \text{ M}$$

if one wishes to study the oxidation of lactate, conditions become more and more favourable the higher the pH. At pH 9 the observed equilibrium constant $K' = 3.7 \times 10^{-3}$ indicates that a starting concentration of 10^{-3} M lactate and 10^{-3} M NAD$^+$ results in a final (equilibrium) concentration of pyruvate and NADH of 6×10^{-5} M each. The rate of appearance of NADH will be perceptibly decreased when the products exceed a concentration of 10^{-6} M. When the appearance of NADH is measured spectrophotometrically such a concentration would result in an absorbancy at 340 mμ of 0.006 in a 1 cm cell. This is the lower limit for an accurate rate measurement with the best equipment. The fluorescence of NADH provides a more sensitive method for the measurement of the rate of appearance of NADH.

It has been shown above that the free energy changes of hydrolysis are often entirely dependent on the subsequent dissociation of products into their ionic forms. Similarly, many oxidative processes are linked to subsequent decarboxylation reactions (p. 54) which make the overall process energetically more favourable than it would otherwise be. The thermodynamic relations of such coupled processes within one enzyme reaction are similar to those discussed in the next section for sequential enzyme reactions.

3.2. SEQUENTIAL ENZYME REACTIONS

3.2.1. The Reactions of Aldolase and of its Products

Muscle aldolase catalyses the reaction

Fructose 1,6-diphosphate $\rightleftharpoons$

glyceraldehyde 3-phosphate + dihydroxyacetone phosphate

and its equilibrium constant at pH 7 and 25° is

$$K = 1.18 \times 10^{-4} \text{ M} \qquad \Delta G° = +5.37 \text{ kcal}$$

It is worth noting that if the reaction is initiated by the addition of fructose diphosphate to a concentration of 1 M the concentrations at equilibrium will be

Fructose diphosphate $\qquad$ 0.9892 M $= C_{FDP}$

Dihydroxyacetone phosphate $\left.\begin{array}{l} \\ \\ \end{array}\right\}$ 1.08×10^{-2} M $= C_{DHAP} = C_{G3P}$
Glyceraldehyde 3-phosphate

giving a ratio of C_{DHAP}/C_{FDP} or C_{G3P}/C_{FDP} equal to 1.09×10^{-2}. If the same calculation is carried out for the initial concentration of FDP as 10^{-3} M, the following results are obtained for equilibrium concentrations:

$$C_{FDP} = 0.711 \times 10^{-3} \text{ M}$$

$$\left.\begin{array}{c} C_{DHAP} \\ C_{G3P} \end{array}\right\} = 0.289 \times 10^{-3} \text{ M}$$

The relative concentrations of DHAP and G3P are now much higher:

$$C_{DHAP}/C_{FDP} \quad \text{and} \quad C_{G3P}/C_{FDP} = 0.406$$

This consequence of the stoichiometry is one of several reasons why the aldolase reaction results in efficient utilization of fructose 1,6-diphosphate in metabolic systems with substrate concentrations of the order of 10^{-3} M. Another factor in the progress of the aldolase reaction is the interconversion of the two products catalysed by triose phosphate isomerase

Glyceraldehyde 3-phosphate $\rightleftharpoons$ dihydroxyacetone phosphate $K = 22$

If an initial concentration of 10^{-3} M FDP is left to equilibrate in the presence of aldolase and triose phosphate isomerase, the final concentrations are:

$$C_{FDP} = 0.45 \times 10^{-3} \text{ M}$$

$$C_{DHAP} = 1.05 \times 10^{-3} \text{ M}$$

$$C_{G3P} = 0.048 \times 10^{-3} \text{ M}$$

3.2.2. General Principles of Consecutive Reactions

The rapid removal of glyceraldehyde 3-phosphate by G3P dehydrogenase and the further rapid removal of the products of this reaction, as well as, in some systems, the removal of DHAP by glycerol phosphate dehydrogenase are also responsible for a large flux through the aldolase reaction. In any process involving two or more steps

$$A \rightleftharpoons B \rightleftharpoons C$$

the equilibria are a function of all the equilibrium constants in the following way:

$$K_1 = \frac{C_B}{C_A} \quad K_2 = \frac{C_C}{C_B} \quad \text{and} \quad K_1 K_2 = \frac{C_C}{C_A}$$

The overall equilibrium constant is the product of all the equilibrium constants of the consecutive steps. This follows also from consideration of the free energy changes. In a multistep process the total free energy change will be the sum of the free energy changes of all the steps:

$$-(\Delta G_1 + \Delta G_2) = RT \ln K_1 + RT \ln K_2$$
$$= RT \ln (K_1 K_2)$$

In a cyclic process

$$\begin{array}{ccc} & K_1 & \\ A & \rightleftharpoons & B \\ K_4 \updownarrow & & \updownarrow K_2 \\ D & \rightleftharpoons & C \\ & K_3 & \end{array}$$

it follows that

(1) at equilibrium all steps must be in equilibrium,
(2) the free energy change on going from A to C must be independent of the route,
(3) $K_1 K_2 = K_3 K_4$

3.2.3. Equilibria between Different Forms of a Substrate

The substrates of many enzyme reactions exist in aqueous solutions in different forms in equilibrium with each other. Examples of such substrates are the α and β forms and open and closed ring forms of sugars, enol and keto forms and free and diol forms of aldehydes. In many cases only one isomer will react with the enzyme. There is some analogy between this phenomenon and the distinction between CO_2 and HCO_3^- as substrate or product of carboxylation and decarboxylation reactions (2.2.6). Some consequences of equilibria between different forms of a substrate or product can be illustrated with the following two examples.

When the oxidation and phosphorylation of glyceraldehyde 3-phosphate

Glyceraldehyde 3-phosphate + phosphate + NAD^+ $\rightleftharpoons$

1,3-diphosphoglycerate + NADH + H^+

is studied in the presence of a moderately high concentration of glyceraldehyde 3-phosphate dehydrogenase the rate of disappearance of G3P can be monitored through the change in extinction at 340 mμ as NADH is produced. As can be seen in Figure 7, the rapid oxidation of a small amount of G3P is followed by a slower steady rate. The small amount which is rapidly removed turns out to be 3 per cent. of the total G3P present, regardless of the absolute concentration. Trentham, McMurray and Pogson (1969) have shown that only 3 per cent. of G3P is in the form of the free aldehyde, the rest is in the hydrated diol form in aqueous solution. The rapid phase of oxidation is the rate of oxidation of free aldehyde, while the slower phase is

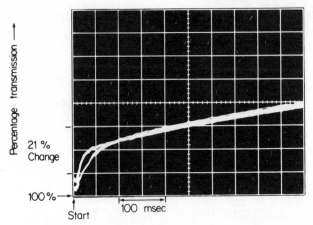

Figure 7 Several superimposed traces from experiments carried out in a stopped-flow spectrophotometer (see Section 8.1.1) record the rate of production of NADH during the reaction of D-glyceraldehyde 3-phosphate and NAD$^+$ with glyceraldehyde 3-phosphate dehydrogenase (in the presence of arsenate as a substitute for phosphate). The substrate concentrations were constant in all experiments and the reaction mixture contained G3P, 0·52 mM; NAD$^+$, 0·48 mM; arsenate, 30 mM; triethanolamine, 0·2 M; pH 8·6 at 22°. The experiments resulting in the faster initial rate were carried out in the presence of 4·0 μM enzyme sites (4/tetrameric protein molecule) and the experiments resulting in slower initial rate were carried out in the presence of 12 μM enzyme sites. After the rapid oxidation of the free aldehyde the rate becomes independent of enzyme concentration and is determined by the rate of dehydration of the diol (see p. 63 and Trentham, McMurray and Pogson, 1969)

determined by the rate of dehydration of the diol:

$$
\begin{array}{ccc}
 & & \text{H} \\
\text{HCO} & \pm\,\text{H}_2\text{O} & \text{HOCOH} \\
| & \overset{k_1}{\underset{k_{-1}}{\rightleftharpoons}} & | \\
\text{HCOH} & & \text{HCOH} \\
| & & | \\
\text{HCOPO}_3\text{H}^- & & \text{HCOPO}_3\text{H}^-
\end{array}
$$

$k_1 = 2\cdot5\ \text{sec}^{-1}$ and $k_{-1} = 8\cdot7 \times 10^{-2}\ \text{sec}^{-1}$ at neutral pH and 20°.

Similar experiments were carried out by Reynolds, Yates and Pogson (1971) on dihydroxyacetone phosphate. The concentration of the free ketone is very temperature-dependent. At 25° it represents 55 per cent.

of the total DHAP present, while at 37° it represents 82 per cent. of the total.
At 25° the reaction

$$
\begin{array}{ccc}
\mathrm{CH_2OH} & \pm\mathrm{H_2O} & \mathrm{CH_2OH} \\
| & \overset{k_1}{\underset{k_{-1}}{\rightleftharpoons}} & | \\
\mathrm{C{=}O} & & \mathrm{HOCOH} \\
| & & | \\
\mathrm{CH_2OPO_3H^-} & & \mathrm{CH_2OPO_3H^-}
\end{array}
$$

has the rate constants $k_1 = 0.21 \ \mathrm{sec}^{-1}$ and $k_{-1} = 0.26 \ \mathrm{sec}^{-1}$.

The fact that only free glyceraldehyde 3-phosphate and dihydroxyacetone phosphate react with the enzymes glyceraldehyde 3-phosphate dehydrogenase, triosephosphate isomerase, aldolase and glycerol phosphate dehydrogenase has a number of implications for the study of enzyme mechanisms and metabolic control. If one wishes to correlate structural changes with binding to enzyme, changes in the proportion of the correct ligand have to be considered. The rapid equilibration of the free aldehydes prior to the slower equilibration of all forms of the substrate may have to be taken into account if transient rather than steady-state fluxes of metabolites occur.

Detailed investigations of the different forms in which substrates can occur and of the enzymatic selection of one of these forms is becoming an important part of the study of many enzyme reactions.

3.3. EFFECTS OF ORGANIZATION ON EQUILIBRIA

3.3.1. Free Energy Changes: Concentration-Dependence

The usefulness of values for standard free energy changes of metabolic processes for the interpretation of real events has at times been over-estimated and may at present suffer the reverse fate. In the previous sections of this chapter the gap between standard conditions and reality of ionic and other environment has been emphasized. The dependence of the interpretation of thermodynamic data on the choice of different pathways is often criticized by purists. Similarly, it is easy to say that standard data are useless for the interpretation of biological systems in the steady state. While steady states are discussed in some detail in Chapter 5, the sections of this chapter are intended to support the view that equilibrium thermodynamics provides a useful framework which one discards at one's peril and overloads with interpretation also at one's peril.

Reactions in cells are rarely in true equilibrium, but they are often not far displaced from it. The free energy change of transferring 1 mole of phosphate from ATP to creatine to form ADP and creatine phosphate is defined under standard conditions, when all four reaction partners are maintained at molar concentration: $\Delta G^{\circ\prime} = +2 \cdot 5$ kcal (at pH 7). At equilibrium when

$$C_{ADP}C_{Creatine\,P}/C_{ATP}C_{Creatine} = 66$$

ΔG for the interconversion is zero. The free energy ΔG at a particular concentration ratio is given by

$$\Delta G = \Delta G^{\circ} - RT \ln \frac{C_{ADP}C_{Creatine\,P}}{C_{ATP}C_{Creatine}}$$

This equation can be applied when one wishes to evaluate the free energy requirements of coupled processes. The relation between free energy changes and absolute concentrations of reactants is an ever recurring theme in a number of sections (1.3.4, 3.1.1 and Chapter 5).

3.3.2. Enzyme–Substrate Compounds and Metabolic Equilibria

The statement that enzymes do not affect equilibria (6.2.4) is correct as long as enzymes are present at very low (catalytic) concentrations. In organized biological systems this condition frequently does not hold. Calculations of the concentration of a particular enzyme in muscle cell sap or intramitochondrial space depend on a number of assumptions. It is, however, generally accepted that many enzymes do occur in these systems to within an order of magnitude of 10^{-4} M and that these concentrations are similar to those of the substrates and of the dissociation constant K_S for

$$ES \rightleftharpoons E + S$$

$$K_S = C_E C_S / C_{ES}$$

These numerical relations have a number of interesting consequences for the behaviour of organized systems. Most intermediates bind to several enzymes: glyceraldehyde 3-phosphate, for instance, will combine with the active site of aldolase, isomerase and dehydrogenase. Most of the substrate is present in ES complexes. Some substrates dissociate slowly from the complex and this dissociation rate controls their availability for other reactions. From the point of view of the present discussion, the most important consequence of compartmentation of substrate by ES compound formation is the fact that the thermodynamic equilibria discussed in previous sections are those of the free forms. Enzyme compound formation involves additional equilibria. In the case of tightly binding substrates like NADH (K Dissociation $\sim 10^{-6}$ M) the presence of E NADH compounds make a considerable contribution to the total NADH concentration. The determination of NADH in a biological system will, by usual methods, result in values for the total (free and enzyme bound) nucleotides. Krebs and his colleagues (for instance, Veech, Raijman and Krebs, 1970) have calculated the ratio of oxidized to reduced nucleotide from values obtained for the ratio of oxidized to reduced substrates of various NAD linked couples. These calculations are based on the fact that the equilibrium of, for instance

$$\text{Lactate} + \text{NAD}^+ \rightleftharpoons \text{Pyruvate} + \text{NADH} + \text{H}^+$$

is dependent on the concentrations of the free nucleotides.

It should, however, be emphasized that enzyme–substrate compounds occur at high concentrations and must be regarded as metabolic intermediates. Although a number of consequences of the significant contribution to the total substrate concentration of these complexes can be predicted, it is not easy to evaluate them quantitatively. The thermodynamic parameters of the formation of the predominant forms of enzyme–substrate complexes are, as yet, rarely known. It is clear that enzymes fulfil other functions in biological systems in addition to the catalysis of the selected reaction. They store metabolites and protect them from reaction with the wrong reagent as well as facilitate the reaction with the correct one.

Two examples of the protective function of enzymes, the flavoproteins and the amino acid activating enzymes, have been discussed by Gutfreund and Knowles (1967) but many others must exist.

CHAPTER 4

Ligand Binding

4.1. LIGAND BINDING TO IDENTICAL NON-INTERACTING SITES

4.1.1. The Applications of Ligand Binding Studies to Enzymes

Most of the topics discussed in Chapter 2 were concerned with the affinity of basic sites for protons. In this connection many fundamental aspects of binding to identical sites were discussed in terms of the thermodynamic parameters of the interaction derived from dissociation constants. In the present chapter the equilibria of ligand binding to large molecules are measured in terms of association constants

$$K = \frac{C_{BL}}{C_B C_L} \tag{4.1}$$

where C_B, C_L and C_{BL} are the molar concentrations of free binding sites, free ligand and ligand-occupied binding sites, respectively. Although the switch from dissociation to association constant is trivial, one is the reciprocal of the other, it is important to get used to the fact that one or other form is used by different authors and that care has to be taken in the interpretation of the literature.

Until the last few years the prime motive for measuring binding constants was to determine the binding energy of a series of related ligands and the number of sites available to the ligand on the macromolecule. This helps to explore the structural complementarity of ligand and binding site. The recent interest in the consequences of ligand binding, the subsequent or associated conformation changes of the macromolecule, has stimulated the development of a wide range of theoretical and experimental procedures for the accurate evaluation of binding energies.

The reason for the emphasis on more accurate analysis, when interaction of binding with other phenomena is to be evaluated, comes from the fact that one starts looking for small perturbations. This will become evident in the next few sections.

The actions of most macromolecular devices (enzymes, carriers or transducers of energy or information) depend on the recognition of specific ligands and the response to interaction with them. In principle it is possible to get all the information about energy changes involved in these processes from kinetic data. If the rate constants of all steps in both directions can be determined by the techniques of Chapter 8, a good deal is learned about mechanism as well as energetics. However, all the necessary kinetic data are rarely obtained and the equilibrium data obtained from ligand binding studies are usually invaluable to complement kinetic information.

A number of examples will illustrate that the algebraic treatment depends on the quantities measured during an experiment. For instance, if some method is available to monitor the free ligand, as in the case of hydrogen ion binding, the following derivation is valuable. Equation (4.1) can be rewritten in the form

$$K = \frac{C_{BL}}{(C_B^\circ - C_{BL})C_L} \qquad \text{where } C_B^\circ \text{ is the total site} \qquad (4.2)$$
$$\text{concentration}$$

$$\frac{C_B^\circ}{C_{BL}} + 1 = \frac{1}{K}\frac{1}{C_L} \qquad (4.3)$$

If C_B° is known and, after each addition of a known amount of ligand, C_L is determined, C_{BL} can be calculated and a plot of C_B°/C_{BL} against $1/C_L$ will have slope $1/K$. The concentration of free hydrogen ions as well as potassium, sodium and an increasing number of other ions can be determined with specific electrodes. If C_B° is not known, it can be determined from the total ligand binding capacity.

The above plot like the Scatchard plot (see Section 4.1.2) and the many algebraic transformations derived from it can be used to give accurate binding constants if they can be shown to result in reliably linear graphs. Linearity is used as evidence for all sites being identical and non-interacting.

4.1.2. The Interpretation of Data

Before attempting to analyse the behaviour of systems giving non-linear plots it is important to discuss the different equations used when different variables are determined experimentally.

As a first experimental method the use of electrodes for the determination of free ligand has been quoted because it was the earliest direct procedure for measuring binding constants. Another group of methods for determining free ligand concentration involves the analysis of the dialysate in equilibrium with the solution. This can be done by ordinary equilibrium dialysis, by running the macromolecular solution through a 'Sephadex' column equilibrated with ligand, by separation of the dialysable fluid through ultrafiltration or ultracentrifugation and by the special techniques of running a solution with increasing ligand concentration through a membrane bounded cell containing macromolecules.

Other methods give direct information about the value for C_{BL}, the concentration of occupied sites. This is the case when the conditions are suitably chosen for experiments with systems in which spectral or fluorescence changes occur when a ligand binding site becomes occupied by a ligand molecule. These changes can be due to perturbation of a chromophore on the ligand or on the macromolecule. For any system in which spectral or fluorescence changes occur these latter methods are usually the preferred ones. If complex phenomena are being studied it is advisable to use two different methods for comparison.

Scatchard (1949), in a classical paper on the interaction of proteins with small molecules and ions, recommends the following derivation. He points out that data should be obtained over the greatest possible concentration range with the greatest possible precision in order that these curves may be extrapolated accurately to both intercepts which provide the required information: the number of binding sites, the value for the binding constant.

From equation (4.2) Scatchard derived

$$\frac{C_{BL}}{C_L} = K(C_B^{\circ} - C_{BL}) \qquad (4.4)$$

On plotting C_{BL}/C_L against C_{BL} a straight line is obtained if K is constant (identical sites and no interaction). The intercept on the C_{BL}/C_L axis (condition $C_{BL} = 0$) gives KC_B°, the binding constant times the number of sites. The intercept at the C_{BL} axis (condition $C_{BL}/C_L = 0$) gives C_B°, the number of binding sites.

For the case of an enzyme molecule with n binding sites the total concentration of binding sites is

$$nC_E^{\circ} = C_B^{\circ} \quad \text{where } C_E^{\circ} \text{ is the total enzyme concentration.}$$

The total ligand concentration C_L° is then given by

$$C_L^{\circ} = C_L + nC_{EL}$$

where C_L is the concentration of free ligand and C_{EL} is the concentration of completely liganded enzyme. If n molecules of enzyme each have, on average, one site occupied this is equivalent to one molecule of enzyme having n sites occupied. R is defined as the fraction of completely liganded enzyme concentration C_{EL}/C_E°, or the fractional saturation. The switch in nomenclature from binding site concentration C_B° to enzyme concentration C_E° is made because often one wishes to determine nC_E° (the total number of sites) when only C_E° is known. In that case one can write

$$K = \frac{nC_{EL}}{nC_E C_L} = \frac{C_{EL}}{(C_E^{\circ} - C_{EL})(C_L^{\circ} - nC_{EL})}$$

On substituting RC_E° for C_{EL} one obtains

$$K = RC_E^\circ/(C_E^\circ - RC_E^\circ)(C_L^\circ - nRC_E^\circ)$$

$$\frac{1}{K} = C_L^\circ\left(\frac{1}{R} - 1\right) - nC_E^\circ(1 - R)$$

$$\frac{1}{K}\frac{1}{1 - R} = \frac{C_L^\circ}{R} - nC_E^\circ \tag{4.5}$$

Figure 8 shows a plot of $1/(1 - R)$ against C_L°/R and the evaluation of n and K.

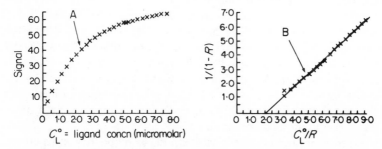

Figure 8   Curve A gives the fluorescence enhancement as NADH is bound to pig heart lactate dehydrogenase. The experiment was performed by Dr. J. J. Holbrook with his differential fluorimetric titrator. A solution of NADH (0·875 mM) was added at a constant rate to a solution of enzyme (2·038 µM in sites) in 0·2 M phosphate buffer pH 7·2 at 25°. A digital print-out records the difference in fluorescence (λ excite = 320 mμ, λ emit = 435 mμ) between the enzyme solution and a buffer solution, to which the same amount of NADH was added automatically. The final difference reading on the scale of the ordinate was 76·446.

Plot B (see equation (4.5)) indicates equivalence and independence of the four sites of this tetrameric enzyme. The dissociation constant calculated from the curve B is 1·064 µM

This type of plot is suitable for the evaluation of the number of sites and the association constant from optical measurements when R can be determined experimentally in the following way, for instance for the case of a spectral change on ligand binding. The ligand is added to two cuvettes (one containing buffer and the other enzyme solution) in a differential spectrophotometer and the change in the extinction difference between the two solutions on adding ligand ($\Delta\Delta\varepsilon$) is measured over a range of total ligand concentration (C_L°) and $\Delta\Delta\varepsilon_\infty$ is determined when further addition of ligand to both cells causes no further change. It should be clear that $R = \Delta\Delta\varepsilon/\Delta\Delta\varepsilon_\infty$. A similar procedure can be used for fluorescence titration of binding sites and this is used for the experiment illustrated in Figure 8. One important assumption is made in a statement above which stipulates that binding n ligand molecules one each to n different enzyme molecules gives the same change as when n ligand molecules bind to one enzyme molecule. When the binding sites are strictly independent this condition should but need not necessarily hold.

4.1.3. Multiple Binding Sites

Although the search for interacting binding sites on oligomeric enzymes has become one of the most fashionable pursuits in enzymology, there are a lot of enzymes with several identical and independent sites per molecule. It is also possible that the sites are independent for one type of ligand while they are cooperative for another type of ligand. Independence versus cooperativity is a topic which will be discussed in Section 4.3 with reference to binding equilibria and in Section 8.3.4 with reference to kinetic analysis. In the present section the main concern is with the evidence for independence and the distribution of ligand among the species of enzyme with one, two, three, etc., ligand molecules bound.

There are a number of numerical methods available to test whether data provide evidence for independence of identical binding sites. None of them are a substitute for accurate data and the best ones depend on data obtained under conditions which are most difficult to study. Weber and Anderson (1965) and Kirtley and Koshland (1967) discuss advantages of various tests for the fit of a single intrinsic association constant to binding data. The most critical test is whether the data obtained at very low saturation $(C_{BL}/C_B^\circ \leqslant 0\cdot1)$ and those obtained near full saturation $(C_{BL}/C_B^\circ \geqslant 0\cdot9)$ can be described by the same association constant K. If $Y = C_{BL}/C_B^\circ$

$$Y = \frac{C_L}{\dfrac{1}{K} + C_L} \tag{4.6}$$

K is the intrinsic association constant, which is the same for all sites on the macromolecule. The ideal titration curve obtained is the same as would be obtained if the same concentration of sites with that affinity were distributed with a single site per macromolecule.

The microscopic association constants calculated from the statistical consideration (see Section 2.2.4) that the first of all sites on one macromolecule has a greater probability of being occupied than the second, and so on, gives for molecules with four sites with intrinsic $K = 1$ for the four microscopic constants:

$$K_1 = 4, \qquad K_2 = 1{\cdot}5, \qquad K_3 = 0{\cdot}67, \qquad K_4 = 0{\cdot}25$$

Clearly the process involves four consecutive equilibria and the overall equilibrium constant $K = K_1 K_2 K_3 K_4 = 1$ as defined above. The practical importance of the statistically corrected individual constants lies in their use for the calculation of the concentrations of the individual species with one, two, three, four sites occupied, respectively.

4.1.4. Thermodynamic Parameters

In spite of the dangers of interpreting thermodynamic parameters in terms of mechanisms, once a method for determining a binding constant has been developed for a particular system it is tempting to determine the temperature dependence and calculate ΔH° and ΔS° for the process. The problems involved are very much the same as those discussed in connection with the mechanism of protein–protein interaction (1.5.3). Changes in solvent structure in the environment of the combining sites tend to over-shadow changes in bond formation between the two interacting molecules.

In connection with discussions of several methods and principles through-out this book the topic of the conformational mobility of protein molecules is an ever-recurring one. The balance of forces near a ligand binding site is bound to be changed when the ligand binds. Several questions arise with respect to this change. Does the ligand tend to stabilize one particular protein conformation out of a large number of possible ones? When a protein molecule exists in a number of equilibrium states, does only one of these bind ligand? It is important to emphasize that even in monomeric enzymes it is of interest to identify the preferred pathway of

$$\begin{array}{ccc} \text{E} + \text{L} & \overset{1}{\rightleftharpoons} & \text{EL} \\ {}^3\Big\updownarrow & & \Big\downarrow{}^2 \\ \hat{\text{E}} + \text{L} & \underset{4}{\rightleftharpoons} & \hat{\text{E}}\text{L} \end{array}$$

to form the complex $\hat{\text{E}}\text{L}$. It is an essential feature of many protein molecules to undergo a rearrangement $\text{E} \rightarrow \hat{\text{E}}$ in response to ligand binding during their function as devices. Thermodynamically sequence 1, 2 is equivalent to sequence 3, 4 (see Section 3.2.2). All that thermodynamics tells one is that the form $\hat{\text{E}}$ is stabilized (has a lower free energy than E) in the presence of ligand. When the subject of allosteric transformations and cooperative changes (4.3.1) first received some publicity, many of these conformation changes of isolated or single units were confused with them (see Rabin, 1966, and Gutfreund and Knowles, 1967).

Mechanisms of conformation changes can be studied in a more meaning-ful manner by the analysis of transient and relaxation kinetic experiments of protein ligand interaction (see Chapter 8). Strictly equilibrium measurements can only distinguish the steps of a two-stage binding process if the methods of observation permit the determination of EL and ÉL individually. Dialysis or other procedures of determining free ligand concentration and total ligand concentration will only provide overall equilibrium information.

4.2. NON-IDENTICAL BINDING SITES

In this section the relatively limited problem of discrete classes of binding sites for one type of ligand on one macromolecule will be discussed very briefly. To some extent the problem is similar to the titration of all the basic groups of a protein. The different classes of binding sites would be the carboxylate ions, the imidazole groups, the α-amino groups and so on. The association constant of a class of ligand binding sites has to be about 100 times larger or smaller than that of the nearest other class of binding sites to make it impossible to obtain distinct constants. This is more pronounced when changes in charge occur at the site during ligand binding. Proton binding gives pronounced interaction between binding sites because of charge effects. For proton and other ion binding phenomena it is not possible to get association without interaction between binding sites. Tanford (1961) and Edsall and Wyman (1958) treat the analysis of proton binding to different classes of sites on a protein molecule in considerable detail.

Figure 9 shows the results of an investigation of the number of binding sites and association constants of the system cytidine triphosphate and aspartate transcarbamylase (see also Section 4.3.5). The Scatchard plot is not linear and two straight lines represent the binding of three ligand molecules each with association constants of $1\cdot0 \times 10^6$ M^{-1} for the first three and $2\cdot1 \times 10^4$ M^{-1} for the next three. The Scatchard plot for a system with two classes of distinct binding sites is described by the equation (see Section 4.1.2)

$$(C_{\text{BL}})_t = (C_{\text{B}}^\circ)_1 \frac{K_1 C_{\text{L}}}{1 + K_1 C_{\text{L}}} + (C_{\text{B}}^\circ)_2 \frac{K_2 C_{\text{L}}}{1 + K_2 C_{\text{L}}} \qquad (4.7)$$

$(C_{\text{BL}})_t$ is the total number of liganded sites at C_{L} free ligand concentration and $(C_{\text{B}}^\circ)_1$ total sites with association constant K_1 and $(C_{\text{B}}^\circ)_2$ total sites with association constant K_2.

There are three models which can produce data of the type shown in Figure 9. The first involves two classes of chemically or structurally distinct sites resulting in intrinsically different binding sites. If the six binding sites are on chemically identical subunits, folding of the polypeptide chain or symmetry arrangements can still result in intrinsically different binding sites. The second possibility is the six intrinsically identical binding sites and the structure of the ligand result in the first three liganded sites interfering

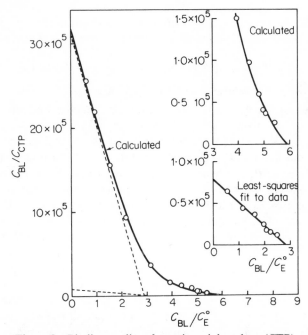

Figure 9 Binding studies of cytosine triphosphate (CTP) to aspartyl transcarbamylase reported by Winlund and Chamberlin (1970). The experiments were carried out by equilibrium dialysis at 4° in 0·01 M potassium phosphate pH 7·0, 2 mM 2-mercaptoethanol and 0·2 mM sodium EDTA. Enzyme was present at 20 mg/ml. For calculated binding constants see p. 70

with binding to the subsequent three sites. This can be a steric or a charge repulsion phenomenon. The third possibility is that the binding of three ligand molecules results in a protein conformation change which decreases the affinity for the remaining sites. The second and third possibilities are special forms of negative cooperativity (see Section 4.3.6). The first possibility can be investigated by a search for selective reactivity and blocking of one or other of the classes of binding sites.

Selective reactivities of sulphydryl groups in aldolase or imidazole groups in ribonuclease in different environments on the same molecule are well known (Gutfreund and McMurray, 1970; Perham and Anderson, 1970; Barnard, 1969). So far there is no good example available for the selective blocking of one class of binding sites for ligands larger than simple ions. As will be seen at various stages during the discussion of ligand binding in more complex systems, the number of proposed model and theoretical treatments by far outnumber the reports of pertinent experimental data.

It will also be seen that equilibrium binding studies are not enough to distinguish between mechanisms; kinetic investigations are also necessary.

4.3. COOPERATIVE BINDING PROCESSES

4.3.1. Historical Considerations of Nomenclature

Cooperative phenomena have been studied by chemists and physicists for a long time and feedback had previously been suggested as an important control mechanism in biological systems when the classical paper by Monod, Changeux and Jacob (1963) focused attention on some problems in enzyme kinetics related to these phenomena. The main feature of this first paper by Monod et al. was to draw attention to the fact that competitive inhibitors need not necessarily compete with the substrate for the same binding site on the enzyme. The mechanism for competitive inhibition had previously always been pictured as competition by a compound of closely related structure. Monod et al. wanted to explain feedback inhibition of the competitive type of one of the enzymes at the beginning of the sequence by the end-product of a chain of enzyme reactions. This end-product competitive inhibitor was no longer '*isosteric*' with the substrate of the inhibited enzyme. It was proposed that these '*allosteric*' inhibitors had their distinct binding sites and that competition between substrate and inhibitor binding occurred through conformation changes of the enzyme in the following way:

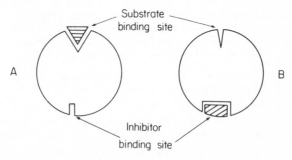

Structure A represents an enzyme molecule with substrate bound and inhibitor site sterically blocked. Structure B represents an enzyme molecule with inhibitor bound and substrate site sterically blocked.

The question whether the free enzyme exists in various forms, only one of which can bind substrate and only one other can bind inhibitor, or whether a two-step mechanism operates with substrate binding prior to the conformation change

$$E + S \rightleftharpoons ES \rightleftharpoons \hat{E}S$$
$$E + I \rightleftharpoons EI \rightleftharpoons \tilde{E}I$$

requires detailed kinetic studies for an answer (see Section 8.3.1). In the case of the two-step mechanism, formation of ESI would be permitted although I cannot combine with $\hat{E}S$ and S cannot combine with $\hat{E}I$.

Allosteric feedback inhibition of enzymes involving structural changes became rapidly confused with any evidence for structure changes during enzyme substrate combination (see 4.1.4). It is also slightly unfortunate that the name for this phenomenon, which does not require the participation of more than one subunit, was carried over when the mechanism was extended to oligomeric enzymes. When this extension occurred the term allosteric was no longer applied to the inhibitor versus substrate structure but became the term for the change in the protein structure when the enzyme changed from the form with the high affinity for substrate to the form with the low affinity for substrate. It must be emphasized that one could get allosteric inhibition without conformation change of the protein and that one does get conformation changes during enzyme–substrate interaction, which have nothing to do with allosteric phenomena. Some aspects of these two possibilities have been discussed in the two previous sections.

Monod, Wyman and Changeux (1965) in a second paper on allosteric phenomena discussed in detail the important control feature of a cooperative switching device introduced in their first paper. Many enzymes occur as aggregates of several identical monomers, some enzymes occur as mixed oligomers of two types of monomeric units. The study of cooperative binding phenomena is concerned with systems in which either substrate or inhibitor or activator binding to one of these units affects the binding of the same compound or of one of the other ones (homotropic or heterotropic interaction, respectively) to other units. As will be seen in different models these phenomena are pictured either as a progressive change, as successive sites are liganded, or an all or none change: with all units in the same one of two conformations. In the second case the equilibrium between the two states will be shifted by progressive liganding.

4.3.2. Homotropic Interaction

The simplest model of cooperativity of interest in enzymology consists of a dimer with two identical monomeric units. Each monomeric unit has one substrate binding site. In response to the first substrate molecule binding on either one of the two identical sites, the affinity for substrate on the other site will increase (in the case of positive cooperativity) or decrease (in the case of negative cooperativity). Homotropic interaction refers to interaction between sites for the same ligand, while heterotropic interaction refers to interaction between sites for different ligands. If the increase in affinity is very large, of the order of 10^3, one has essentially the situation that the concentration of ES will be negligible compared with E and ES_2. As a result of this the normally hyperbolic plot of R against C_S becomes sigmoidal. In Figure 10 R is the fraction of the total number of binding sites which is occupied by substrate, while C_S is the free substrate concentration. R is also C_{ES_2}/C_E° (cf. Section 4.1.2).

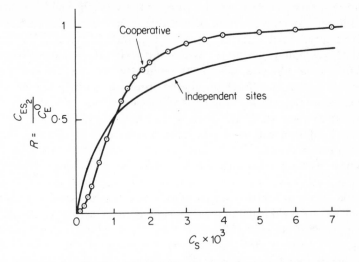

Figure 10   Simulated hyperbolic (for independent sites) and sigmoidal (for cooperative interaction) titration curve

For complete cooperativity, when only free enzyme and enzyme combined with two substrate molecules occur, one writes the association constant

$$K = \frac{C_{ES_2}}{C_E C_S^2}$$

$$\log K + 2 \log C_S = \log \frac{C_{ES_2}}{C_E^{\circ} - C_{ES_2}}$$

In the general case of strongly cooperative binding sites of an enzyme with n sites the equation becomes

$$\log K + n \log C_S = \log y \qquad (4.8)$$

where

$$y = \frac{C_{ES_n}}{C_E^{\circ} - C_{ES_n}} = \frac{R}{1 - R} \qquad (4.9)$$

A plot of $\log y$ against $\log C_S$ has a slope n and intercept $\log K$. This is called the Hill plot, n is called the Hill constant and equation (4.8) is the Hill equation. The maximum value for n is the number of binding sites per enzyme molecule. In practice Hill's n is usually smaller than the number of binding sites and it gives a measure of the cooperativity. In the rest of this chapter the relevant properties of a few selected proteins and their interactions with specific ligands will be discussed. These examples have been chosen to illustrate in some detail the different physical concepts of models for cooperative binding phenomena. The current theories used will be summarized in the final section.

4.3.3. Oxygen Binding to Myoglobin and Haemoglobin

Both myoglobin and haemoglobin have the physiological function to carry oxygen, albeit under different conditions. The cooperative properties and the linkage between oxygen and carbon dioxide binding to haemoglobin provide the most extensively studied molecular regulation of a biological process. Myoglobin is a monomeric protein with one haem group, while mammalian haemoglobins are tetrameric proteins with four haem groups— one attached to each monomer. Haemoglobin carries oxygen from the lung to various tissues and carries CO_2 back from the tissues to the lung. Myoglobin provides an oxygen store in muscle.

The haem group on each of the monomers ($M = 17,000$), the single one of myoglobin or the four making up haemoglobin, is an iron II complex of protoporphyrin. The iron remains in the ferrous form during oxygen binding and release. The whole porphyrin ring is buried in a hydrophobic cleft of the protein and this environment prevents the oxidation Fe^{2+} to Fe^{3+}. Free haem is rapidly oxidized to the ferric state in the presence of oxygen.

The single polypeptide chain of myoglobin and the two different (two α and two β) chains which make up the tetrameric haemoglobin, have very similar structures. The specific difference in the behaviour of the different chains must be due to differences in some of the finer detail of the structure. The differences in the structure of these three monomeric units have expressed themselves in their ability to interact. Myoglobin as well as a preparation of pure α-chain of haemoglobin remain monomeric under a wide range of conditions. The isolated β-chain of haemoglobin forms tetramers. An equimolar mixture of α- and β-chains forms tetramers of the $2\alpha2\beta$ type. The tetrameric haemoglobin will dissociate reversibly into $\alpha\beta$ dimers at low protein and/or high salt concentrations. There is a marked increase in the tetramer–dimer dissociation constant on binding oxygen to haemoglobin.

Several other ligands (CO, NO, alkylisocyanides) also bind reversibly to the Fe^{2+} of haemoglobin and myoglobin. Various ligands have been used for different experiments; their quantitatively different behaviour makes each of them more suitable for the exploration of different features of the behaviour of respiratory proteins. Qualitatively the processes are very similar with the various ligands.

Figure 11 shows several features of the behaviour of the haemoglobin–oxygen system. Myoglobin combines with oxygen with a single association constant of 10^6 M. With haemoglobin the sigmoidal plot of R versus oxygen pressure indicates a change in association constant as the four sites are

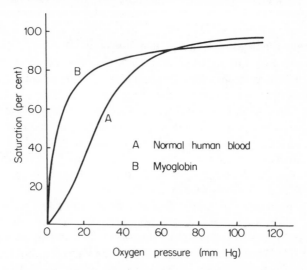

Figure 11   The oxygen dissociation curves for myo-globin and haemoglobin illustrate that oxygen will be transferred to myoglobin under conditions of low oxygen tension. The increase in oxygen affinity of haemoglobin, as the oxygen tension increases, is also apparent

progressively liganded. The low affinity of haemoglobin for oxygen at low oxygen pressure results in the transfer of oxygen from haemoglobin to myo-globin. Adair (1925) derived a general equation for oxygen binding to the four sites with four different progressively larger binding constants:

$$R = \frac{A_1 C_{O_2} + 2A_2 C_{O_2}^2 + 3A_3 C_{O_2}^2 + 4A_4 C_{O_2}^4}{4(1 + A_1 C_{O_2} + A_2 C_{O_2}^2 + A_3 C_{O_2}^3 + A_4 C_{O_2}^4)} \tag{4.10}$$

The use of As for the composite constants of the general Adair equation was proposed by Roughton in honour of Adair. This equation was intended to describe the oxygen saturation curve of haemoglobin regardless whether the four intrinsic association constants are identical or whether they are non-equivalent. If the four Ks are intrinsically identical this means that in the unliganded tetramer all four sites have an equal probability of receiving the first ligand molecule and of the remaining three sites all would have an equal probability of receiving the second ligand molecule and so on. However, once the first site is liganded interaction can change the probability of any of the next three sites accepting a ligand and so on. The equation can then be written in the form

$$R = \frac{K_1 C_{O_2} + 2K_1 K_2 C_{O_2}^2 + 3K_1 K_2 K_3 C_{O_2}^3 + 4K_1 K_2 K_3 K_4 C_{O_2}^4}{4(1 + K_1 C_{O_2} + K_1 K_2 C_{O_2}^2 + K_1 K_2 K_3 C_{O_2}^3 + K_1 K_2 K_3 K_4 C_{O_2}^4)} \qquad (4.11)$$

If the Ks are intrinsically identical K's and there were no interaction to change them as they are progressively liganded then $K'_1 = K_1/4$, $K'_2 = 2K_2/3$, $K'_3 = 3K_3/2$, $K'_4 = 4K_4$ (see 2.2.4 and 4.1.3). These statistical corrections have to be added to any progressive change due to interaction changes during liganding of the sites. At very low C_{O_2} equation (4.11) tends to the limiting form

$$R = K_1 C_{O_2}/4$$

At very high C_{O_2} the equation tends to a limiting form which permits evaluation of K_4. Roughton, Otis and Lyster (1955) have given examples of the procedures developed in Roughton's laboratory for the evaluation of four constants which describe the whole oxygen binding curve. There are an enormous number of different sets of values obtained by different authors on haemoglobins from different species under different conditions. K_1, K_2 and K_3 are usually within a factor of three of each other, while K_4 is between 20 to 70 times larger than the first three. Without kinetic information it is not possible to propose a mechanism for the cooperative binding of oxygen to haemoglobin. However, the extension of a proposal made by Roughton that a conformation change subsequent to combination with four oxygen molecules results in a decrease in the rate of dissociation from haemoglobin fits many of the facts. An alternative hypothesis which fits some of the experimental evidence depends on a change of structure when two molecules of oxygen are bound, resulting in the remaining two sites being more reactive. One of the many points which have not been settled is whether the α- and β-chains of unliganded haemoglobin have intrinsically different affinities for ligand.

More and more detail of the three-dimensional structure of liganded and unliganded haemoglobin is becoming available through the work of Perutz and his colleagues (1970). The description of the structural and chemical work is out of place here but two more topics should be summarized briefly.

The thermodynamic aspects of the linked equilibria of CO_2 and O_2 binding to haemoglobin have been described by Wyman (1964). The Bohr effect (Bohr et al., 1904) is one of the classical observations on haemoglobin. Perutz has shown that oxygen binding to each subunit causes structural changes in the protein which result in changes of pKs and the availability of amino groups for carbamate formation (see Section 2.3.6). Unliganded haemoglobin binds CO_2 and through the uptake of protons from the solvent also makes CO_2 (in the form of HCO_3^-) more soluble. Figure 12 illustrates the well-established interaction between CO_2 and the oxygen affinity of haemoglobin. The lowering of oxygen affinity is partly due to combination of CO_2 with the amino terminal groups of the constrained (unliganded) structure.

The dissociation of haemoglobin into dimers in dilute solution and/or at high salt concentration is also linked to the structural change caused by ligand binding. The conformation of the unliganded tetramer results in firmer association between the two $\alpha\beta$ dimers than is the case in the liganded conformation. This tetramer–dimer dissociation is not significant under physiological conditions, when the haemoglobin concentration is very high indeed, but it provides a useful tool for the study of some properties of this protein. Kellett and Gutfreund (1970) showed that kinetic studies of the association of dimers to tetramers, after rapid removal of ligand, can be used to determine the contribution of ligand binding to dimer–tetramer interaction. From Perutz's (1970) model and kinetic studies it is clear that in the unliganded conformation there are several salt bridges between subunits, additional to those observed in fully liganded haemoglobin. Further kinetic studies showed that the degree of dissociation into $\alpha\beta$ dimers in partially liganded haemoglobin solution is a measure of the transition between the two forms of low and high affinity for ligand. The dimer is always in the form with high affinity for ligand. The whole tetramer structure is required for the cooperative structure changes between the two forms.

4.3.4. Homotropic and Heterotropic Interactions in a Dimer

The shortage of examples of equilibrium binding data makes it necessary to use data obtained by an indirect method to illustrate some phenomena. If the velocity v of an enzyme reaction in the steady state, at a specified substrate concentration, is used as a measure of the concentration C_{ES} of the enzyme–substrate complex, then one can write

$$\frac{C_{ES}}{C_E^\circ - C_{ES}} = \frac{v}{V - v} \qquad (4.12)$$

where V corresponds to the maximum velocity obtained at saturating substrate concentration.

The validity of this method is discussed in detail in Section 6.3.1. While in many cases the substrate concentration dependence of the reaction velocity is only a function of the substrate affinity, in some cases the rate constants of catalytic steps also enter into the equations for the substrate

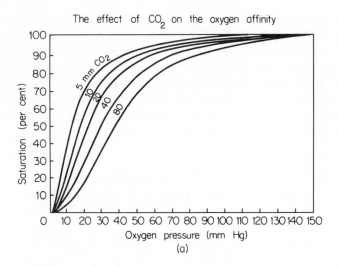

The effect of CO_2 on the oxygen affinity

(a)

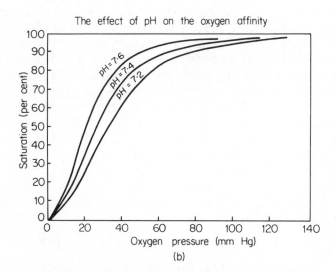

The effect of pH on the oxygen affinity

(b)

Figure 12 The Bohr effect, the lowering of oxygen affinity, by increasing CO_2 concentration, is illustrated by the two sets of curves. As oxygen is removed and haemoglobin goes into the constrained structure, the CO_2 carrying capacity is increased. CO_2 combines with the free amino endgroups of haemoglobin and the liberation of protons, as the structure of haemoglobin changes from the liganded to the unliganded (constrained) conformation, results in an increased solubility of CO_2

saturation curve. In the latter case equation (4.12) will not describe substrate binding curve correctly. Only the direct determination of the rate constants of enzyme–substrate complex formation by transient and relaxation techniques (Chapter 8) can give kinetic information about equilibrium binding.

A well-documented example of homotropic and heterotropic interaction of ligand binding on a dimeric enzyme is provided by the studies of Hess and his colleagues on yeast pyruvate kinase. The binding parameters were obtained from measurements of the enzyme catalysed reaction

$$\text{Phosphoenol pyruvate} + \text{ADP} \rightleftharpoons \text{Pyruvate} + \text{ATP}$$

Figure 13 shows the results of a number of experiments performed by Hess, Haekel and Brand (1966) in which they studied the cooperativity of substrate (PEP) and fructose diphosphate (activator) binding. ATP (a feedback inhibitor) also binds cooperatively.

A model can be proposed which requires two forms of enzyme: the T-form and the R-form. The T-form has a low affinity for PEP and a high affinity for ATP, which results in stabilization by ATP binding of this form and inhibition of the reaction at a fixed PEP concentration. The R-form has a high affinity for PEP and FDP, which results in FDP stabilization of this form and activation of the reaction at a fixed PEP concentration.

The homotropic interaction of the two PEP binding sites on the molecule is demonstrated by a Hill plot of data obtained for varied PEP concentration in the absence of ATP and FDP. The slope of n was nearly 2. The binding constant of PEP to the T- and R-forms of the enzyme can be computed in a variety of ways. In the absence of FDP or ATP the affinity calculated from the experimental points obtained at very low PEP and very high PEP concentrations will respectively give K_{PEP}^{T} and K_{PEP}^{R}. Also the association constants calculated from the curves in Figure 13 should give at high ATP concentration K_{PEP}^{T} and at high FDP concentration K_{PEP}^{R}. The theoretical limiting curves for both cases have been plotted by the computer. The ratio of the affinities c is given by

$$c = K_{PEP}^{T}/K_{PEP}^{R} = 1 \cdot 9 \times 10^{-4}/4 \cdot 8 \times 10^{-2} = 0 \cdot 004$$

Another important constant of such systems is L_0 the equilibrium constant of the transformation $R_0 \rightleftharpoons T_0$ of the free unliganded enzyme. For yeast pyruvate kinase $L_0 = 250$ and this constant is called the allosteric constant in the nomenclature of Monod et al. The significance of L_0 in relation to particular allosteric models will become apparent later.

The phenomenon of heterotropic interaction is illustrated by the effects of ATP and of FDP binding on the affinity for PEP. These different ligands each have separate sets of binding sites but binding of one of them has an effect on the other.

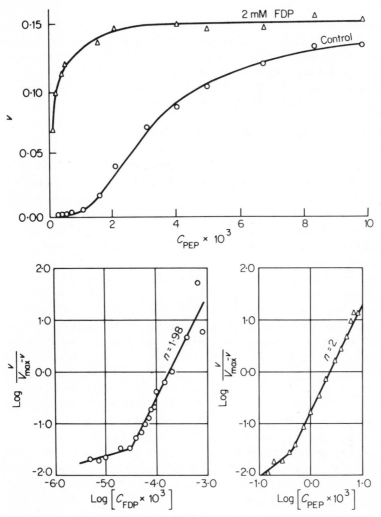

Figure 13   The experiments of response of yeast pyruvate kinase activity to phosphoenol pyruvate and fructose diphosphate concentration (from Hess et al., 1966) are discussed on p. 84

Hess and his colleagues have also shown that the transition from the R-form to the T-form is slow compared with the turnover of the enzyme-catalyzed reaction. Studies of rates of conformational transitions will be discussed in Sections 8.1.4 and 8.3.4, but it is important to mention at this stage one distinction between kinetic and equilibrium effects of cooperativity. This effect is important for enzymes and not for carriers like haemoglobin.

When the data derived from steady state rate measurements are used for plots of the type shown in Figure 13 one does not necessarily distinguish between cooperative effects on binding and cooperative effects on catalysis. While the effects on binding are better documented in the literature, it is also possible to imagine a system which would change into one with greater catalytic activity when the sites are progressively occupied.

4.3.5. Polymers, Ligands and Conformational Transitions

So far cooperative changes in a dimer and in a tetramer have been discussed. If one now goes on to consider hexamers, the inevitable question arises whether the number of units in a polymeric enzyme has to be an even one. There is a long history of investigations which first produced evidence for trimeric enzymes, which turned out to be tetrameric on more careful examination. Most methods are likely to err on the side of giving too few subunits or too few sites. All one can say at this stage is that there is no well-documented case of a trimer or pentamer, but this does not prove that they do not exist. As will be seen (4.3.6), one of the major theories of cooperative transitions in biological control relies on symmetry, but with the limited evidence available one should beware of such generalizations. In large aggregates the number of subunits cannot be estimated with an accuracy of ± 1. Cooperative phenomena are also postulated to operate as switching devices in membranes: contact of a few sites with a chemical messenger could transform the properties of a whole array of molecules.

In the case of haemoglobin the tetramer consists of two types of monomers. Although all four monomers bind O_2, the α- and β-units differ in some of their properties. Heterogeneity in subunit structure is more marked in some enzymes which have binding sites for one kind of ligand on one type of subunit and sites for another type of ligand on other subunits. A typical example of such a system is aspartyl transcarbamylase, ATCase, the enzyme which catalyses the first step in the biosynthesis of pyrimidine in *E. coli*. The reaction catalysed by this enzyme is

L-aspartate + carbamyl phosphate $\rightleftharpoons$ carbamyl L-aspartate + phosphate

and the ultimate product of the pathway, cytidine triphosphate, is a strong inhibitor of ATCase. This inhibition is characteristic for the type of feedback phenomenon discussed in 4.3.1.

ATCase has a molecular weight of 310,000 and it consists of six subunits of $M = 33,000$ and six subunits of $M = 17,000$ (Wiley and Lipscomb, 1968 and Weber, 1968). It was found in the earlier studies of Gerhart and Pardee (1962) that one type of unit ($M = 33,000$) binds the substrates and the other type the feedback inhibitor. The inhibitor sites can be inactivated without affecting the catalytic ones and it was also found possible to separate the two types of subunits with their binding sites intact. An experiment which demonstrates that the six inhibitor sites may not be identical is illustrated in Section 4.2 (Figure 9).

The available data at the time of writing are not suitable for the quantitative evaluation of cooperativity of ATCase. This enzyme can, however, be used to illustrate an interesting phenomenon already mentioned. As in the case of haemoglobin, one can study the binding properties of the two types of sub-units. Although the isolated catalytic or inhibitor subunits form polymers in those systems there is no interaction between binding sites. In the complete enzyme system with six catalytic and six inhibitor (regulatory) subunits there is distinct cooperativity in succinate (a substrate analog) binding.

Other enzymes are more suitable for the demonstration of cooperative binding (see Section 4.3.3 and below) but another important aspect of co-operativity can be illustrated well with ATCase. It is most important to find out more about the molecular mechanism of cooperativity than can be learned from equilibrium binding studies. Kinetic investigations into the mechanism are discussed in Section 8.3.3. In this section some attention will now be given to the evidence for structure changes which can be obtained from changes in reactivity and spectra of the protein.

Gutfreund and McMurray (1970) have recently reviewed the wide range of reactivities found for sulphydryl groups in proteins and their sensitivity to changes in environment. Gerhard and Schachman (1968) have examined the reactivity of the sulphydryl groups of ATCase and its two components in isolation and the effects of ligands on these reactivities. These phenomena can now be studied much more effectively and quantitatively. It should be possible to determine the rate of conformation changes, as has been attempted for haemoglobin (see Antonini and Brunori, 1970). In haemoglobin a particular SH group (93 in the β-chain) is approximately 30 times more reactive in liganded as compared with unliganded haemoglobin. When haemoglobin was mixed simultaneously with oxygen and with mercurial it was, however, found that even in a flow apparatus (see p. 178) with a time resolution of 10^{-3} sec it was not possible to measure the delay in the activation of the SH reactivity after oxygen binding. The conformation change resulting in the cooperative behaviour is very fast indeed. However, it turns out that the time constants of conformational transitions in different cooperative phenomena vary over at least as wide a range as from about 10^{-4} to 10^2 sec.

In the case of aspartokinase-homoserine dehydrogenase, which catalyses the reactions

ATP + L-aspartate $\rightleftharpoons$ ADP + 4-phospho-L-aspartate

L-aspartate β semialdehyde + NADH + H$^+$ $\rightleftharpoons$ L-homoserine + NAD$^+$

the conformational transition of the enzyme on ligand binding can be observed through spectral changes of the protein. This enzyme is in control of a key reaction in the biosynthesis of amino acids from aspartate in *E. coli.* The hexamer consists of three each of two slightly different, but largely homologous, subunits. Each subunit has a molecular weight of 60,000. Preoccupation with the interesting cooperative properties has caused some neglect in the study of the equally unusual catalytic properties of this and some other regulatory enzymes. The exploration of the relation between the two catalytic activities and the binding sites on the six subunits may give a lot of information on the regulatory phenomena.

There are only three binding sites for NAD^+ or NADH and only three catalytic sites on the hexamer. On the other hand, the cooperative feedback inhibition by threonine of both aspartokinase and dehydrogenase activity operates through six inhibitor binding sites. The enzyme is only sensitive to full regulatory effects of threonine in the presence of K^+ ions. The specific effects of small ions on the structure and activity of enzymes are worthy of detailed study for an understanding of many control phenomena.

Janin and Cohen (1969) found that aspartokinase-homoserine dehydrogenase has the interesting property that large changes in the absorption spectrum (in the region 250 to 310 mμ) and in the fluorescence of the protein occur when various ligands are added in the presence of 0·15 M KCl. It was therefore possible to study the equilibrium between different conformational states by optical methods (see Figure 14).

From experiments of this type the following conclusions could be drawn. The optical properties indicate that there are only two conformations T and R. The T-form is favoured by binding of threonine and has about 15 per cent. of the maximum homoserine dehydrogenase activity but does not have aspartokinase activity (does not bind substrate). The R-form is favoured by binding of K^+ and aspartate and has both enzyme activities.

The effect of temperature on the equilibrium between the two forms of the enzyme was also studied by Janin and Cohen. They found that $\Delta H° = 28$ kcal for the conversion of R $\rightarrow$ T indicates a large increase in enthalpy during the process, which involves removal of exposed tyrosine and tryptophan residues into a more compact structure.

Another phenomenon which is linked to ligand binding is a change in the degree of polymerization of proteins. Again the amount of quantitative information available is not sufficient to warrant a detailed account. However, this is likely to be a problem of much future interest in connection with the study of the assembly of polymers. Studies of systems which have been investigated indicate that changes in interaction between monomeric units may only be the consequence of the important structural changes and not the essential part of the control phenomenon. For instance, Kellett (1971) has shown that the tetramer $\rightarrow$ dimer dissociation of haemoglobin is

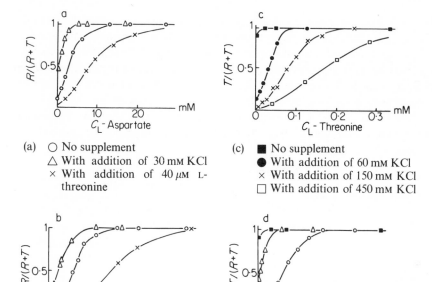

(a) ○ No supplement
     △ With addition of 30 mM KCl
     × With addition of 40 μM L-
       threonine

(c) ■ No supplement
     ● With addition of 60 mM KCl
     × With addition of 150 mM KCl
     □ With addition of 450 mM KCl

(b) ○ No supplement
     △ With addition of 2 mM L-
       aspartate
     × With addition of 40 μM L-
       threonine

(d) △ With addition of 5 mM L-
       aspartate
     ○ With addition of 20 mM L-
       aspartate

Figure 14 The fraction of aspartokinase-homoserine dehydrogenase in the
R-state $[R/(R + T)]$ and the fraction in the T-state $[T/(R + T)]$ can be determined
from the extent of the change in protein fluorescence on addition of different
ligand (Janin and Cohen, 1969)

markedly increased on oxygen binding. However, under physiological
conditions in the red blood cell neither the liganded nor the unliganded
structure of haemoglobin will dissociate into the dimeric form to any extent.

Frieden (1971) has recently surveyed the effects of ligand binding on
protein–protein interaction. The polymerization and activity of glutamate
dehydrogenase is a system which has been studied extensively. This enzyme
is made up from polypeptide chains of molecular weight 50,000. Each of
these chains has one active site but dissociation of the protein into units of
$M = 50,000$ in special solvents results in loss of activity. The smallest active
molecule appears to be a hexamer. The active molecule, $M = 300,000$,
undergoes further polymerization to aggregates of molecular weights larger
than 2×10^6. At one time it was thought that the specific activity of enzyme
with $M = 2 \times 10^6$ was higher than that of the hexameric enzyme with

$M = 300,000$. It was found by Frieden that this was due to the fact that some inhibitors bind more strongly to the hexamer than to the larger polymers.

Figure 15 shows Scatchard plots of the binding of the inhibitor guanosine triphosphate to glutamate dehydrogenase at different concentrations of the enzyme. The state of aggregation of the hexamer into larger units at different concentrations is also indicated.

Several other enzymes undergo readily reversible association–dissociation phenomena into species with different affinities for effectors and different

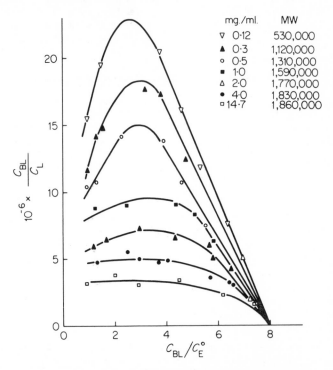

Figure 15   This study of GTP (guanosine triphosphate) binding to glutamate dehydrogenase (beef liver) is reproduced from Frieden and Colman (1967). In this diagram the values for C_{BL} were calculated from the number of moles of GTP bound per mole of enzyme (based on a molecular weight of 400,000). C_L represents the concentration of free GTP. The extrapolated value of 8 sites per molecule should be corrected to 6 sites if the presently preferred molecular weight of 330,000 is accepted. The experiments were carried out at 5° in 0·1 M tris-acetate buffer, pH 7·1, with 1 mM phosphate, 1 mM EDTA and 400 μM NADH

activities. The two muscle enzymes phosphofructokinase and phosphorylase are extensively studied examples. The data available so far are, however, not sufficient to document a new principle for the present discussion.

4.3.6. Some Theories of the Mechanism of Interaction between Binding Sites

The general Adair equation (4.3.3) does not suggest a mechanism but provides a means of describing any binding curve by a set of constants. Monod, Wyman and Changeux (1965) stimulated the field by proposing a very specific mechanism for cooperative interaction between substrate and effector binding sites. Their model is illustrated in Figure 16 for the case of a tetramer. Their proposal stipulates that all subunits are either in the T-form □ with low affinity for ligand or all are in the R form ○ with high affinity for the ligand. As one, two, three and four ligand molecules are bound the more will the R-conformation for the whole tetramer be favoured. The

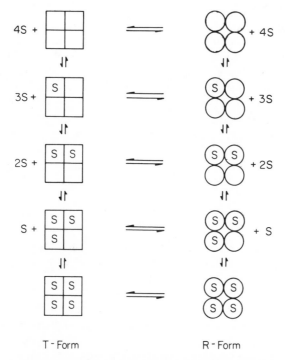

T - Form   R - Form

Figure 16 The 'all or none' model of Monod, Wyman and Changeux (1965) in which all the units of a polymer have to be the same form, T or R. Progressive liganding favours the R form, which has the greater affinity for ligand

equilibrium constant $L = C_T/C_R$ will depend on the number of ligand molecules bound and there will be L_0, L_1, L_2, L_3, L_5 for the five rows in the scheme in Figure 16. However, there will be only two ligand-binding constants—one for binding to the T- and one to the R-form of tetramer. The simplicity of this model is appealing.

Koshland, Nemethy and Filmer (1966) have criticized the model of Monod et al. and have set up much more general equations for the interaction between sites. The principle of the method of Koshland et al. can be illustrated with the scheme presented in Figure 17. If the sequence of ligation the tetramer is diagonal and the intermediate forms

exist, as well as the completed T- and R-form, five distinct binding constants could be found.

As Eigen (1968) has pointed out, it is impossible to get convincing evidence for any scheme without kinetic studies of the rates of transition between the different forms. The differences between the two idealized models are discussed further in Section 8.3.3. The difficulty is that the degeneracy of the scheme (absence of intermediate states) is as much a function of the sensitivity of the experimental methods used as it is of the actual physical model.

The model of Monod et al. cannot account for negative cooperativity—the decrease in binding constant as successive sites are ligated. Accurate studies of individual binding constants could exclude the all-or-none model if negative interaction is indicated. Symmetry relations between binding constants

$$K_1 K_4 = K_2 K_3$$

should apply to the Koshland model but need not apply to the Monod model. It must be pointed out that the existence of only two forms to which ligand can bind in the Monod model does not exclude the possibility of fitting four Ks to a titration curve. This is due to the fact that L_0, L_1, L_2, L_3 and L_4 will change the apparent Ks during the progress of ligation. With the enormous variety of functions required from polymeric molecules in biological systems, it would be very surprising if it did not turn out that different mechanisms function for different purposes. In Section 8.3.3 kinetic studies of such systems are reported.

Information about the molecular mechanism of the interaction can also be obtained from studies which reveal the thermodynamic parameters of the process. It is found that the Hill n of the haemoglobin–oxygen reaction is

T - Form   R - Form

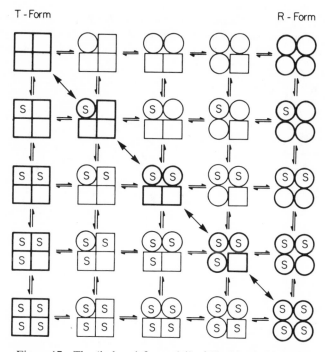

Figure 17   The 'induced fit model' of Koshland, Nemethy
and Filmer (1966) in which the units of a polymer can individually change from the T-form into the R-form as the
polymer is progressively liganded. In its idealized form
progress in the above scheme would be diagonally across
different intermediates as the polymer is progressively
liganded. In practice the finer detail of the mechanism can be
studied as outlined in Section 8.3.3

almost temperature independent. This means that the free energy of interaction is largely due to an entropy change. It has been mentioned that the
Hill constant is a measure of the energy involved in the cooperative interaction. It now remains to show how this can be evaluated from experimental
data.

Often one finds Hill plots presented as straight lines with slope n, but it is
difficult to determine the points representing low concentrations of ligand.
On theoretical grounds one must obtain a curve of the type shown in Figure
18 if accurate data can be obtained at limiting low and limiting high ligand
concentrations. At these limits, when only the first or last site is being
liganded, the slope of the plot must be 1. The distance between the two
asymptotes is a function of the free energy of interaction. Wyman (1968)

treats this subject in some detail. If there is no interaction the Hill plot will have unit slope throughout. If at any point the binding curve is steeper than for the case of a simple titration curve, at that point (log C_L) the slope of the

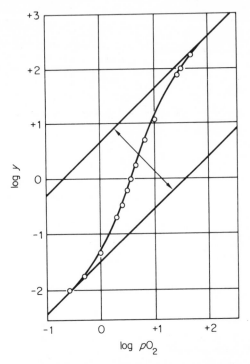

Figure 18   The Hill plot of oxygen binding to sheep haemoglobin in 0·2 M phosphate buffer at pH 9·1 and 19°. The data of Roughton, Otis and Lyster (1955) have been plotted on a scale which demonstrates the asymptotic approach to unit slope by Wyman (1968). The free oxygen concentration is expressed as partial pressure in mm Hg and y is defined in equation (4.9)

Hill plot will be steeper than unit slope. Graphically the integrated interaction energy is given by $RT\sqrt{2}$ times the distance between the two asymptotes. This type of calculation only gives the correct answer if the binding sites are initially identical. Calculations with this assumption give a free energy of interaction for oxygen binding to haemoglobin of approximately 3 kcal per site.

CHAPTER 5

Transport Processes

5.1. FROM TIME INDEPENDENCE TO TIME DEPENDENCE

5.1.1. Introducing Non-equilibrium Thermodynamics

In previous chapters we have been concerned with time-independent processes: forces or potentials at equilibrium. The introduction of time as an additional variable in considerations of distribution of energy or distribution among states should form a bridge between thermodynamics and kinetics. In biological systems, just as in a chemical engineering process, equilibria are the end of the road: the process is finished, the cell is dead. Equilibrium thermodynamics is an essential tool in the laboratory for the calculation of the fundamental properties or potentialities of substances or systems of substances. It now remains to be seen how the lessons learned from observations of systems at equilibrium can be applied to the interpretation of the dynamic behaviour of these systems.

There have been quite a number of discussions of the possibility that biological systems might not obey the second law of thermodynamics. Some of these were due to a misunderstanding or wrong application of the unqualified statement by Clausius: 'The energy of the World is constant, the entropy of the World tends towards a maximum'. The spontaneous increases in complexity associated with the evolution of life were considered a possible contradiction of the law of increase in entropy. Other discussions of the relation between entropy, order and information contents of biological systems have emphasized the oversimplification of the generalizations due to the direct use of these terms for the description of biological phenomena (see Medawar, 1969, for a review).

Here we are concerned with the proof that the physical laws hold for any system as long as they are applied correctly. The correct application depends on a precise definition of the system and the properties of the walls round it. Parts of this chapter are based on the excellent treatment of non-equilibrium thermodynamics by Katchalsky and Curran (1965).

5.1.2. Systems

A system is a geometrically defined part of the universe on which we wish to perform an experiment. The properties of the wall which isolates the system from the rest of the universe, or from another circumscribed system, distinguish three types of systems:

An adiabatic system is enclosed by a wall which is impermeable to heat or matter.

A closed system is enclosed by a wall which is permeable to heat but impermeable to matter.

An open system is a geometrically defined space, which can freely exchange heat and matter with its surroundings.

While any process occurring in an adiabatic system results in an increase in entropy, in an open system the influx of energy or matter from its environment can maintain processes which result in a decrease in entropy. In biology we encounter systems of open systems and each of these collections of systems is again only an open system and so on ad infinitum. The truth of Clausius' statement applied to the universe (rather than to the World) depends on whether the universe is an adiabatic system; the Earth is certainly not.

If we have a small open system within a larger adiabatic system and a process within the small system results in a decrease in entropy, the increase in entropy of the larger system must be at least equal to the decrease in the smaller one. Much of non-equilibrium thermodynamics is concerned with coupling of processes occurring either in two connected systems or within a non-homogeneous system. It will be relatively easy to demonstrate how systems and processes cooperate in a qualitative manner. It is, however, difficult to provide quantitative proof without serious assumptions inherent in the application of thermodynamic concepts to non-equilibrium processes.

5.1.3. Steady-states

When a system is not in equilibrium it is easiest to interpret or predict its behaviour if it is in a steady-state or at least close enough to it to be correctly described by the steady-state approximation. It is important that the conditions of steady-state approximations are properly understood also for the correct use of many solutions of kinetic problems (see Section 6.3.4).

Steady-states are sometimes referred to as dynamic equilibria because the concentrations of intermediates are constant. In a sequence of reactions:

$$A + B \xrightarrow{k_1} \boxed{X \xrightarrow{k_2} Y} \xrightarrow{k_3} C$$

X and Y will reach a constant steady-state level if the concentrations of A and B are so large that during the course of the observation the change in their concentrations is infinitesimally small compared with their total concentrations. A and B are regarded as a constant source and C as a sink.

It follows that $C_A C_B k_1, k_2$ and k_3 are all constant and that C_X will be constant when

$$dC_X/dt = k_1 C_A C_B - k_2 C_X = 0$$

$$C_X = k_1 C_A C_B/k_2$$

C_Y will be constant when

$$dC_Y/dt = k_2 C_X - k_3 C_Y = 0$$

$$C_X/C_Y = k_3/k_2$$

There is a continuous turnover of X and Y even when C_X and C_Y are constant and the flow of molecules through the constant pool of X and Y is called the chemical flux.

Perfect steady-states are not likely to be found in real systems and it is important to give some more thought to the steady-state approximation mentioned above. The source concentration is not likely to be perfectly constant and therefore the size of the X and Y pool is not likely to be perfectly constant. For the moment we are neglecting another possible perturbation of the pool size, namely the effects of the concentration of C in the sink. If the latter changes this can affect the rate of $Y \rightarrow C$ in a variety of ways. To take into account that no steady-state is perfect but that the steady-state approximation is valid, one says that in the steady-state the rate of change in the intermediate concentrations (dC_X/dt and dC_Y/dt) is negligibly small compared with the rate of flux through these intermediates (dC_C/dt). It is, of course, purely arbitrary what one calls negligible in a theoretical statement like this. In practice one has the test that during the time of observation the concentrations of the intermediates do not change by a larger amount than the accuracy of the experimental measurement of any of the concentrations (see Section 6.3.2).

5.2. APPROACH TO EQUILIBRIUM

5.2.1. Flux and Flow

Another set of approximations, which are essential for the theory of non-equilibrium thermodynamics used at present both for transport processes and for the analysis of chemical relaxations (see Section 8.3.1) is the assumption that any perturbation from equilibrium is sufficiently small to be reversible and linear in concentration. Some idea as to how small sufficiently small is should be gained when these approximations are applied below.

Steady-state or transport processes are often described in terms of flux; chemical flux is described either in the terms used above or in a simple process of $A \rightarrow B$ in terms of flux through a transient intermediate or an energy barrier (see Section 7.1.3). In general one can describe flux by saying that the flow of 'something' (flux through a plane perpendicular to the direction of flow) is proportional to the gradient of this 'something' at that plane. The rate of the chemical reaction $A \rightarrow B$ is proportional to C_A. This is discussed in some detail in the next section. Similarly the flow of heat is proportional to the temperature gradient, the flow of electricity is proportional to the gradient of electric potential and diffusion of a solute is proportional to the concentration gradient of the solute across the boundary. A general relation between flux J and the driving force X through a coefficient L for the particular phenomenon (chemical reaction, diffusion, etc.) is

$$J = LX \tag{5.1}$$

The physical significance of J, X and L is best illustrated with two processes: chemical flux during a reaction and mass flow during diffusion.

5.2.2. Chemical Flux

The force driving a chemical reaction is the affinity A which is defined by

$$A = -\sum v_i \mu_i \tag{5.2}$$

v_i is the stoichiometric coefficient and μ_i the chemical potential of reactant species i. The expression is summed over all reactants.

A comparison of the kinetic with the thermodynamic derivations for chemical flux in the reaction

$$A \underset{k_{-1}}{\overset{k_1}{\rightleftharpoons}} B$$

will illustrate some of the crucial assumptions of non-equilibrium thermodynamics. At the same time we shall derive J, X and L for this process.

First the kinetic approach. At any one time the rate of change in the concentration of reactant A is given by

$$dC_A/dt = -k_1 C_A + k_{-1} C_B \tag{5.3a}$$

and the corresponding change of the concentration of B is

$$dC_B/dt = +k_1 C_A - k_{-1} C_B \tag{5.3b}$$

the flux through the chemical barrier is

$$J = -dC_A/dt = +dC_B/dt \tag{5.4}$$

At equilibrium the net transport through the barrier $J = 0$. Therefore,

$$k_1 \bar{C}_A = k_{-1} \bar{C}_B \tag{5.5}$$

where $\bar{C}_A$ and $\bar{C}_B$ are the equilibrium concentrations. The equilibrium constant for the reaction is

$$K = \bar{C}_B/\bar{C}_A = k_1/k_{-1}$$

and deviation of the actual concentrations from the equilibrium concentrations are defined by

$$\alpha_A = C_A - \bar{C}_A \tag{5.6a}$$

$$\alpha_B = C_B - \bar{C}_B \tag{5.6b}$$

The assumption, referred to above, that only processes very close to equilibrium are considered is defined by the statement that

$$\alpha_A/\bar{C}_A \ll 1 \quad \text{and} \quad \alpha_B/\bar{C}_B \ll 1 \tag{5.7}$$

A condition which is introduced to reduce the number of variables is the conservation of matter, no exchange of reactants with the surroundings:

$$C_A + C_B = \bar{C}_A + \bar{C}_B$$

and

$$\alpha_A + \alpha_B = 0 \tag{5.8}$$

If flux is expressed as (see equation (5.4))

$$J = k_1(\bar{C}_A + \alpha_A) - k_{-1}(\bar{C}_B + \alpha_B)$$

and this is simplified through introduction of

$$k_1\bar{C}_A - k_{-1}\bar{C}_B = 0 \qquad \text{(from equation (5.5))}$$

and

$$\alpha_B/\alpha_A = -1 \qquad \text{(from equation (5.8))}$$

Therefore,

$$J = \alpha_A(k_1 + k_{-1}) = k_{-1}\alpha_A(1 + K) \tag{5.9}$$

For the thermodynamic approach to the derivation of an expression for J the relation between affinity and chemical potential is used. As stated above in the case of a chemical process the force X is the affinity A and

$$A = \mu_A - \mu_B$$

This simple expression for the chemical potential difference between the two reactants at C_A and C_B is applicable in a unimolecular process when the stoichiometric coefficient is unity. The equilibrium condition is that the chemical potentials of the two reactants are equal:

$$\bar{\mu}_A = \bar{\mu}_B \tag{5.10}$$

when the concentrations of the reactants are C_A and C_B. The chemical flux expressed in terms of the general relation (equation (5.1)) is

$$J = XL = (\mu_A - \mu_B)L \tag{5.11}$$

The chemical potential difference can be transformed into a function of the concentrations through the relations between standard chemical potentials for the two reactants and the chemical potentials at concentrations in the reaction mixture (see Sections 1.3.2 and 1.3.3)

$$\mu_A = \mu_A^0 + RT \ln C_A \tag{5.12}$$

It will be seen that the following rearrangement of the term $RT \ln C_A$ is very useful:

$$\underline{RT \ln C_A} = RT \ln (\bar{C}_A + \alpha_A)$$

$$= RT \ln (\bar{C}_A + \alpha_A) \frac{\bar{C}_A}{\bar{C}_A}$$

$$= RT \ln \left(1 + \frac{\alpha_A}{\bar{C}_A}\right) \bar{C}_A$$

$$= RT \ln \bar{C}_A + RT \ln \left(1 + \frac{\alpha_A}{\bar{C}_A}\right) \tag{5.13}$$

and the affinity can be written as

$$A = \mu_A^0 + RT \ln \bar{C}_A + RT \ln \left(1 + \frac{\alpha_A}{\bar{C}_A}\right)$$

$$- \mu_B^0 - RT \ln \bar{C}_B - RT \ln \left(1 + \frac{\alpha_B}{\bar{C}_B}\right) \tag{5.14}$$

This is readily simplified since

$$\bar{\mu}_A = \bar{\mu}_B = \mu_A^0 + RT \ln \bar{C}_A = \mu_B^0 + RT \ln \bar{C}_B$$

(see equations (5.10) and (5.12)). Therefore,

$$A = RT \left[\ln \left(1 + \frac{\alpha_A}{\bar{C}_A}\right) - \ln \left(1 + \frac{\alpha_B}{\bar{C}_B}\right) \right]$$

The terms in the form $\ln [1 + (\alpha_A/\bar{C}_A)]$ can be simplified when the condition $\alpha_A/\bar{C}_A \ll 1$ holds. In such a case the first term of the series

$$\ln \left(1 + \frac{\alpha_A}{\bar{C}_A}\right) = \sum_{n=1}^{\infty} (-1)^{n-1} \left(\frac{\alpha_A}{\bar{C}_A}\right)^n n^{-1}$$

can be used for the approximation

$$\ln\left(1 + \frac{\alpha_A}{\overline{C}_A}\right) \approx \frac{\alpha_A}{\overline{C}_A}$$

Therefore,

$$A = RT(\alpha_A/\overline{C}_A - \alpha_B/\overline{C}_B)$$

and

$$A = RT\frac{\alpha_A}{\overline{C}_A}(1 + K) \tag{5.16}$$

this gives the thermodynamic expression

$$J = RT\frac{\alpha_A}{\overline{C}_A}(1 + K)L \tag{5.17}$$

from the above and the kinetic expression (equation (5.9))

$$J = k_{-1}\alpha_A(1 + K)$$

the phenomenological coefficient L relating the force to the flux of a chemical process is

$$L = k_{-1}\overline{C}_A/RT \tag{5.18}$$

This linear relation between flux and concentrations of reactants is clearly only applicable to very small deviations from chemical equilibrium.

5.3. DIFFUSION AND SEDIMENTATION AND SOME PROPERTIES OF MOLECULES

5.3.1. Isothermal Diffusion

An understanding of models and rates of diffusion processes is important for the interpretation of many biological phenomena as well as for the characterization of the size and shapes of solute molecules. Diffusion in an isothermal system in the absence of any coupled chemical or physical process is called free isothermal diffusion. Once the free diffusion rates of a solute are established the coupling of diffusion to other phenomena can be investigated.

Again kinetic and thermodynamic derivations can be obtained and the phenomenological equation for the flux due to diffusion can be derived from them. For simplicity diffusion will be considered first in one dimension

(x) only. The solution is considered to be homogeneous in the other two dimensions (y and z):

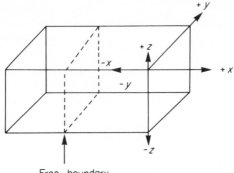

Free boundary

The columns of the two solutions are assumed to be long enough in the x direction for the concentrations of all components to remain constant at the end of the columns during the course of the experiment. This corresponds to an infinite source and an infinite sink.

The kinetic argument used to derive equations for diffusion processes comes from the random movement of molecules. In the y and z dimension the number of solute molecules moving in the + and − direction will, on average, be equal. However, with the solution on the left-hand side of the boundary and pure solvent on the far right of the boundary, more solute molecules will be moving in the +x compared with −x direction through the boundary. This results in mass movement of solute towards the pure solvent. Fick's first law describes this phenomenon:

$$J = -D\left(\frac{\partial C_S}{\partial x}\right)_t \tag{5.19}$$

At a given time t the flux at a point x will be dependent on the gradient of the solute concentration (C_S) and the proportionality constant or diffusion constant D.

The thermodynamic argument derives from the fact that the force responsible for diffusion is the difference in chemical potential of the solute (μ_S) on the two sides of the boundary. The system will change towards equilibrium, which is reached when the chemical potential of all components is identical throughout. The phenomenological equation is then written as

$$J = -L\left(\frac{\partial \mu_S}{\partial x}\right)_t \tag{5.20}$$

Using the relation $\mu_S = \mu_S^\circ + RT \ln C_S$ (see Section 1.3.2) one can combine Fick's equation with the phenomenological equation. However, the constant of practical importance is the diffusion constant, which can be evaluated from observations of the change in the concentration with time. As will be summarized below, the phenomenological coefficient is required for the theoretical analysis of the interaction of a number of non-equilibrium processes occurring in the same system.

5.3.2. The Evaluation and Physical Significance of the Diffusion Constant

The units of flux in a mass diffusion process are moles cm^{-2} sec^{-1}, which gives the amount of solute crossing unit area in unit time. The concentration gradient has the units moles cm^{-3} cm^{-1}, i.e. moles cm^{-4}. From equation (5.19) one can deduce that the units of the diffusion constant D are cm^2 sec^{-1}.

The velocity V_S in the one dimension, the direction of the potential energy gradient grad μ_S is given by

$$V_S = \frac{1}{f} \operatorname{grad} \mu_S \qquad (5.21)$$

where f is the frictional resistance to the movement of the solute molecules. In dilute ideal solution this can be written in terms of concentration

$$V_S = -\frac{RT}{fC_S} \operatorname{grad} C_S$$

Since $J = V_S C_S$ it follows that

$$J = \frac{RT}{f} \operatorname{grad} C_S$$

and from this and equation (5.19) it follows that

$$D = \frac{RT}{f} \qquad (5.22)$$

The frictional resistance is given by Stokes as

$$f = 6\pi r \eta_0 \qquad (5.23)$$

where r is the radius of the particles and η_0 is the viscosity of the particle moving at unit velocity through pure solvent. The units of viscosity are dyne sec cm^{-2} and the units of RT are erg $mole^{-1}$ or dyne cm $mole^{-1}$. The friction F for one mole is Nf, where N is Avogadro's number. Hence

$$D = \frac{\text{Dyne cm mole}^{-1}}{\text{Dyne sec cm}^{-1}\text{mole}^{-1}} = cm^2 \, sec^{-1}$$

For the measurement of diffusion one starts off ideally with a perfectly sharp boundary with pure solvent on one side and homogeneous solution on the other side. The random movement of solute molecules results in symmetrical boundary spreading increasing with time.

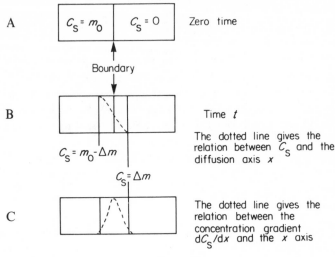

Figure 19

The dotted line in Figure 19 C is a Gaussian distribution curve and the concentration m_x at point x and time t is

$$m = \frac{m_0}{2} - \frac{m_0}{\sqrt{\pi}} \int_0^y e^{-y^2}\, dy \qquad (5.24)$$

where $y = x/2\sqrt{(DT)}$.

This is the solution of Fick's second equation

$$\left(\frac{\partial m}{\partial t}\right)_x = D\left(\frac{\partial^2 m}{\partial x^2}\right)_t \qquad (5.25)$$

The most common method for measuring diffusion constants is to observe the change in distribution of solute across the boundary by an optical method. Either the refractive index or the absorbancy of the solution is measured as a function of distance from x_0 at different times and $(\partial m/\partial x)_t$ is recorded. The derivative

$$\left(\frac{\partial m}{\partial x}\right)_t = \frac{m_0}{2\sqrt{(\pi Dt)}} e^{-x^2/4Dt} \qquad (5.26)$$

will have the value $m_0/2\sqrt{(\pi Dt)}$ at $x = 0$.

In practice there are a great variety of more or less accurate methods for the determination of diffusion constants (Tanford, 1961; Schachman, 1957). The evaluation and use of diffusion constants for the determination of molecular weight and shape will be referred to again at the end of the next section.

The development of lasers has provided the possibility of extending the range of information which can be obtained from the study of light scattering. The narrow bandwidth of the laser beam permits the analysis of fluctuations in solutions due to transport and reactions.

5.3.3. Sedimentation Velocity and Equilibrium

Transport under the influence of gravitational forces is of considerable practical interest for the study of properties of molecules in solution. If a particle in solution is subjected to a force $\mathscr{F}$ in a centrifugal (gravitational) field, it will accelerate until the steady-state velocity is reached. The steady-state velocity dx/dt is determined by a balance between the forces applied to the particle by gravitation and by friction. The frictional resistance per particle was defined in 5.3.2 as $f(dx/dt)$ and it follows that the steady-state rate of sedimentation dx/dt will occur when

$$\mathscr{F} = f\frac{dx}{dt} \quad \text{and} \quad \frac{dx}{dt} = \frac{\mathscr{F}}{f}$$

The gravitational force acting in the centrifuge on a particle of mass m and density $1/\bar{v}$ (where $\bar{v}$ is the partial specific volume, or volume in cm^3 taken up by 1 g) in a medium of density ρ is

$$\mathscr{F} = \omega^2 xm(1 - \bar{v}\rho) \tag{5.27}$$

ω is the angular velocity and x the distance from the centre of the rotor.

High-speed centrifuges are used which run at speeds of about 1,000 revolutions per second and with distances of about 6 cm between the centre of rotation and the solution. The angular velocity

$$\omega = \text{revolutions sec}^{-1} \, 2\pi \, (\text{radians});$$

at 1,000 revolutions sec^{-1} and 6 cm radius of rotation

$$\omega^2 x = (6{,}280)^2 \times 6 = 2{\cdot}366 \times 10^8 \, \text{cm sec}^{-2}$$

While the above value is needed for further calculations of sedimentation velocities, a comment should be made on the evaluation of the commonly used relation between centrifugation and normal gravity. Acceleration due to gravity on Earth is $981 \, \text{cm sec}^{-2}$ and the above conditions would be referred to as $2{\cdot}4 \times 10^5 \times g$.

If the mass of the particle under examination is defined by

$m = M/N$ (M is molecular weight and N Avogadro's number)

we obtain

$$\mathrm{d}x/\mathrm{d}t = [\omega^2 x M(1 - \bar{v}\rho)]/Nf$$

The sedimentation constant S is defined as

$$S = (\mathrm{d}x/\mathrm{d}t)/\omega^2 x$$

and it follows that S is related to the molecular weight of the sedimenting particle by

$$S = [M(1 - \bar{v}\rho)]/Nf \tag{5.28}$$

If one wishes to calculate the molecular weight of a molecule from measurements of the sedimentation velocity one needs to know $\bar{v}$, ρ and f. The first two of these parameters are determined by accurate density measurements. The frictional constant f is an important parameter characterizing the properties of a molecule in solution. The diffusion constant in terms of N particles

$$D = RT/Nf$$

can be combined with the sedimentation constant to determine both M and f.

$$M = RTS/D(1 - \bar{v}\rho) \tag{5.29}$$

If one uses the molecular weight of a molecule to calculate the frictional constant for a molecule as if it were a solid (unsolvated) perfect sphere one obtains f_0. The frictional ratio f/f_0, where f is evaluated from experimental data, provides a measure of the asymmetry and solvation of the molecule under investigation. Wyman and Ingals (1943) have constructed very useful nomograms of the interrelation of shape, hydration, frictional ratio and measured hydrodynamic quantities. The variety of contributions of properties of macromolecules to their sedimentation and diffusion constants has made it difficult to obtain absolute values for asymmetry or hydration.. It is, however, now possible to make very accurate differential measurements of sedimentation constants of a protein under several conditions and to detect changes in conformation introduced by these conditions (Kirschner and Schachman, 1971).

The major inaccuracy of calculating molecular weights from equation (5.29) comes from small uncertainties about $\bar{v}$. The diffusion and sedimentation constant is usually quoted as D_{20}° and S_{20}°, corrected to the value of water at 20° (Schachman, 1957) which makes RT in equation (5.29) ($8\cdot314 \times 10^7$ ergs/mole) 293°K. From sedimentation alone the molecular weight of a globular protein can be estimated approximately on the assumption that $\bar{v} = 0.75$ and $f/f_0 = 1.2$ (a reasonable value)

$$M = 5{,}378 \times S_{20}^{2/3} \times 1.2$$

Similarly the rate of diffusion of globular proteins can be calculated from the molecular weight on the assumption that the value for f/f_0 is about 1.2 times that calculated from the radius for a particle of that molecular weight.

Several of the techniques used to determine the size of particles in solutions involve the observation of the approach to equilibrium between diffusion and sedimentation. Katchalsky and Curran (1965) derive the equations describing the behaviour of such systems. They in fact start with the description of equilibria and then use the concept of flux to introduce steady-state sedimentation.

If one starts with a homogeneous system of a protein and buffer solution in an ultracentrifuge cell and subjects this to a centrifugal field, in due course a boundary will form with only buffer solution on one side and protein in buffer on the other side. For simplicity it is assumed for a first approximation that the centrifugal field is not strong enough for significant effects on the homogeneity of the buffer salt concentration throughout the cell. As the boundary is formed diffusion across it will occur. In a sedimentation velocity experiment the speed of the centrifuge is high enough to make the movement of the boundary sufficiently fast for its spreading through diffusion to be minimal. If the centrifugal field is reduced and sedimentation slowed down one can find a proper range of speeds at which an equilibrium distribution of protein concentration will eventually be established throughout the cell.

The distribution of chemical potential of the protein component (μ_P) throughout the cell will be

$$-d\mu_P/dx = -d\mu_P^C + M_P(1 - \bar{v}_P\rho)\omega^2 x \tag{5.30}$$

where $d\mu_P^C$ is the concentration gradient contribution to the chemical potential gradient and M_P and $\bar{v}_P$ refer to the molecular weight and partial specific volume of the protein. When equilibrium is reached the chemical potential must be constant throughout the cell and it follows that

$$(\mu_P^C)_{x_1} - (\mu_P^C)_{x_2} = RT \ln \frac{(C_P)_{x_2}}{(C_P)_{x_1}} = M_P(1 - \bar{v}\rho)\frac{x_2^2 - x_1^2}{2}$$

$$M_P = \frac{2RT \ln [(C_P)_{x_2}/(C_P)_{x_1}]}{\omega^2(1 - \bar{v}\rho)(x_2^2 - x_1^2)} \tag{5.31}$$

where x_1 and x_2 refer to distances from the centre of rotation. Sedimentation equilibrium studies (like osmotic pressure) can be used to determine the concentration dependence of the chemical potential of protein molecules and thus to evaluate molecular interaction (5.4.2).

5.4. OSMOSIS, INTERACTIONS AND ELECTROKINETIC PHENOMENA

5.4.1. Osmosis

Osmosis, like diffusion, is a process of importance for the physical maintenance of biological systems and also the basis of a method for studying properties of systems of large molecules. While in diffusion one is concerned with the flow of solute, in osmosis one is concerned with the flow of solvent.

The simplest model for osmosis consists of two compartments with pure solvent in one and a solution of large molecules in the same solvent in the other compartment. The two compartments are separated by a membrane which is freely permeable to the solvent and impermeable to the large solute molecules. The solvent in contact through the membrane will tend towards the condition of equilibrium

$$\mu^{I} = \mu^{II}$$

At equilibrium the chemical potential of component I (the solvent) must be the same in the pure solvent compartment I and the solution compartment II. The solute has no communication between the two compartments and cannot equilibrate.

If the temperature and pressure are the same in the two compartments the difference in the solvent chemical potential is

$$\mu_1^{I} - \mu_1^{II} = RT \ln \frac{C_1^{II}}{C_1^{I}}$$

For equilibration the solvent concentration would have to be the same on both sides. In an attempt to achieve this, solvent will flow from the solvent compartment into the solution compartment until the solution is infinitely diluted—this flow is osmosis.

When the solution compartment is enclosed or provided with an expansion tube pressure will build up in compartment II as solvent flows in from compartment I. This pressure will increase the chemical potential of the solvent in compartment II (see Section 1.3.2) and equilibrium will be reached before $C_1^{II} = C_1^{I}$—this pressure is called the osmotic pressure.

5.4.2. Osmotic Pressure, Molecular Weights and Interactions

The osmotic pressure of a solution is due to the mole fraction of the solvent being reduced in the solution compartment by the solute. The pressure is therefore proportional to the mole fraction of the solute (see, for instance, Gutfreund, 1950). At 4° the osmotic pressure is related to the molecular weight of the solute by

$$P = C_2^{II}RT = (g/M)\,22{,}400$$

where P is the pressure in cm of water, g the concentration of the solute in g/l. and M the molecular weight of the solute. The proportionality of the osmotic pressure to the molar concentration of the solute only holds in ideal solutions. Deviations from ideality provide information about properties of the solute in the particular solution studied.

When one considers enzymes or other charged macromolecules one introduces the complication of charge interaction which can make a significant contribution to the activity of protein molecules. The conditions of osmotic pressure measurements are chosen either to simulate the natural environment, if the chemical potential in vivo is to be determined, or one uses buffers and neutral salts to minimize charge interaction, if the molecular weight is to be determined.

Osmotic pressure measurements are valuable for the study of subunit dissociation phenomena. In this connection it is important that in heterogeneous systems of, for instance, a mixture of dimers and tetramers, different methods provide fundamentally different information about molecular weights. The combination of sedimentation and diffusion data for such a system in dynamic equilibrium will give a weight average molecular weight $\overline{M}_W$

$$\overline{M}_W = \sum g_i M_i / \sum g_i \qquad (5.32)$$

while osmotic pressure and sedimentation equilibrium give the number average molecular weight $\overline{M}_N$

$$\overline{M}_N = \sum C_i M_i / \sum C_i \qquad (5.33)$$

In the above expressions g_i is the concentration in g/l. and C_i is the molar concentration of the component with molecular weight M_i.

5.4.3. Some Electrokinetic Phenomena

In Section 1.4.3 the relation between the size, charge and mobility of ions was briefly discussed. At first sight the problem of deriving equations for the movement of ions under the influence of an electric field is very similar to that involved in describing sedimentation. The electromotive force acting on charged particles makes them accelerate until the force equals $(dx/dt)f$, the frictional resistance. This is why in a comparable series of ions as listed on p. 18 the mobility is inversely proportional to the frictional radius. Conversely, if one keeps the frictional radius of a molecule constant one can relate its charge to the electrophoretic mobility. In this way the isoelectric pH and the number and sign of the charges on an enzyme molecule can be evaluated from electrophoresis.

The complexities of electrochemical investigations are due to the strong interactions between the various ions which take place in such systems. The topics mentioned in this section are only discussed very briefly for the sake of providing something approaching a complete list of the physical phenomena contributing to transport processes. For a significant treatment of these topics and their application to biophysical systems, Katchalsky and Curran (1965) provide an excellent introduction and further references.

Membrane potentials and electro-osmotic phenomena are going to become increasingly important concepts for the interpretation of the behaviour of enzyme systems in compartments or bound to membranes. If solute molecules, like proteins, are charged and enclosed by a membrane which is impermeable to proteins but permeable to water and smaller ions, the distribution of smaller ions will be affected by the excess charge in the protein compartment. The consequent unequal distribution of positive and negative diffusible ions will both contribute to the osmotic pressure of the protein solution and create an electric potential across the membrane—the membrane potential. The significance of this phenomenon in the interpretation of osmotic pressure measurements is reviewed by Gutfreund (1950) and the importance of membrane potentials for processes linked to electron and ion transport is discussed by Mitchell (1966). This phenomenon is also responsible for liquid junction potentials.

Electric potentials are also formed across a diffusion boundary of electrolytes–diffusion potentials. These are due to the difference in size and hence rate of diffusion of the two partners of an ion pair. A balance will be set up between the energy required to maintain this potential and the separation due to diffusion.

Electro-osmosis is the flow of solvent due to an electromotive force at constant pressure. The direction of flow will depend on the charge of the membrane and the direction of the potential. These and other electrokinetic phenomena are discussed in detail in physiology texts. They are of greater importance in the study of more organized systems than those of concern to enzymology at this time. There is no doubt, however, that greater organization of molecular assemblies will become of more and more interest to the molecular enzymologist.

5.5. TRANSPORT AND CARRIERS

5.5.1. Some Definitions

The simplest method of transporting metabolites from one part of the cell to another or through a membrane is free diffusion. The laws governing free diffusion were discussed in Section 5.3.1. The adjective 'free' implies that there is no control or selectivity in this process apart from the chemical potential gradient of the diffusing substance.

The next level of organization of transport across a membrane is found in cases of selective permeability. Some organization is inherent in the separation of different locations by a membrane which is impermeable to enzymes. This results in some reactions taking place on one side of the membrane and the products react further on the other side of the membrane. The situation becomes more interesting when the metabolites do not diffuse freely through the membrane but are selectively translocated.

Most metabolites are ionic and are not readily miscible with the lipid component of the membrane. However, combination with a specific hydrophobic protein (carrier) of the membrane can facilitate translocation (see also Figure 20 for a low molecular weight carrier). Facilitated transport of this type is characterized by selection of the compounds which can pass

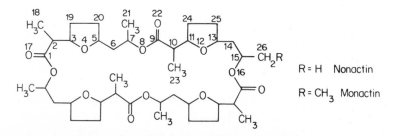

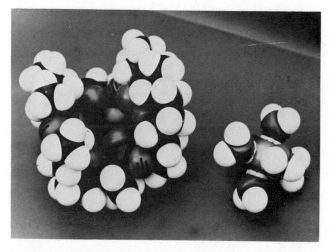

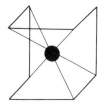

Figure 20   Many antibiotics act as specific carriers for ions. From structure analyses and kinetic investigations the above model for complex formation has been postulated. This model involves the dehydration of the ion during combination with the carrier and could well be the basis of many translocation mechanisms (Diebler, Eigen, Ilgenfritz, Maas and Winkler, 1969)

through the membrane and by a limited concentration dependence. While free diffusion is strictly concentration-dependent, carrier-facilitated diffusion reaches a saturation rate when all sites on carrier proteins are occupied: the rate-limiting step is translocation after the metabolite binds to the carrier. The carrier can also be a freely soluble compound which, together with the 'passenger', forms a lipid soluble compound for translocation. The compound illustrated in Figure 20 are examples of carriers for small ions.

An increasing number of proteins is being isolated from the membranes and the periplasmic space of bacteria. These are of relatively small size ($M \sim 30,000$) and bind certain metabolites very specifically and very tightly (for a recent review see Kabak, 1970). It is not yet known whether some of these proteins act as specific sites or carriers for translocation of the type discussed above or whether they are always involved in active transport.

Active transport is defined as movement of substances against a concentration gradient coupled to a chemical process (5.5.3). Transport against a concentration gradient can also be coupled to another diffusion process in the opposite direction. The basic requirement for a system of coupled processes to proceed is that the sum of all the free energy changes of the component processes is negative and the sum of the entropy changes is positive. The dissipation of free energy per unit time Φ (see Katchalsky and Curran, 1965; Katchalsky and Spangler, 1968)

$$\Phi = T\frac{dS}{dt} = \sum_i J_i X_i \geqslant 0 \qquad (5.34)$$

This defines the conditions for the sum of the individual fluxes due to different processes having conjugate forces X driving them. Many physical principles can be involved in the coupling of such flows. Here one gets involved in the design of molecular devices, in the relation between molecular recognition, information and entropy and the control of permeability by electric potentials. Hypothetical models for such coupled processes are discussed in the next two sections.

5.5.2. Symport and Antiport

A carrier-controlled transport system across a membrane can be designed to work *against* the concentration gradient of one substance provided it works at the same time with the concentration gradient of the other. This type of coupling is easily visualized in terms of the coupling of fluxes expressed in equation (5.34).

If a carrier X combines both with glucose G and with Na^+, but it has to bind these sequentially,

$$G + X \rightleftharpoons XG$$

$$Na^+ + XG \rightleftharpoons XGNa^+$$

and if glucose binds weakly on its own but very tightly once Na$^+$ is bound, one can visualize the following system:

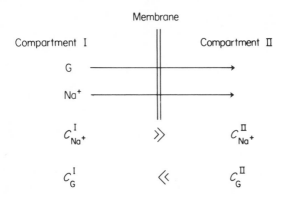

Only carrier in all its three forms (free, with glucose, with glucose and Na$^+$) can go through the membrane and will tend to be at or near equilibrium throughout. When carrier combines with glucose in compartment I there is a large probability that at the high Na$^+$ concentration XGNa$^+$ will be formed and diffuse through to compartment II. At the low concentration of Na$^+$ in compartment II both ligands will dissociate from the carrier and the free carrier will diffuse back to compartment I. The efficiency of the process depends on the relation

$$K_G = \frac{C_{XG}}{C_X C_G} \qquad K_{GNa^+} = \frac{C_{XGNa^+}}{C_{XG} C_{Na^+}}$$

and

$$\frac{C_G^I}{C_G^{II}} \bigg/ \frac{C_{Na^+}^{II}}{C_{Na^+}^I}$$

From values for this concentration ratio one can calculate

$$\{C_{XG}^I + C_{XGNa^+}^I\}/\{C_{XG}^{II} + C_{XGNa^+}^{II}\}$$

and the rate of transport of glucose from compartment I to II is proportional to this ratio. This phenomenon is called symport or parallel transport.

Another model involves cross-transport or antiport and involves coupling of flow of a metabolite against the concentration gradient to flow of another substance in the opposite direction with its concentration gradient. Supposing a carrier can bind both Na$^+$ and K$^+$ and only carrier and carrier bound ions can diffuse through the membrane between the two compartments:

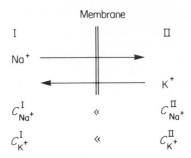

Under such conditions XK^+ will diffuse from compartment II to I and dissociate to X plus K^+. Free X will then combine to some extent with Na^+ (depending on the relative binding constants and concentrations of the ions) and XNa^+ will diffuse to compartment II where exchange of XNa^+ to XK^+ occurs. The efficiency of the process will again depend on a relation between the ratios of the concentrations in the two compartments and the carrier binding constants.

The question which might be asked is: what is the relation between enzyme mechanisms and carrier mediated transport mechanisms? The emphasis on recognition and on the response of enzyme molecules to initial contact with specific ligand molecules goes through much of the discussion in Chapter 4. It has also been pointed out that small ions have significant effects on substrate binding and the conformation changes of enzymes associated with catalysis and control. The study of the mechanism involved in the functions of carriers and especially of carrier proteins involves very much the same kind of equilibrium and kinetic techniques as investigations of enzyme substrate interaction. The relation between free and membrane-bound carrier proteins has its analogy in the study of enzymes. The changes in the activity and specificity of some enzymes, when they are in homogeneous dispersion or membrane bound, are among the exciting borderlines between molecular enzymology and the study of more organized biological systems.

5.5.3. Active Transport

Active transport has been defined as movement against a concentration gradient coupled to a chemical reaction. Some mechanisms simply involve a chemical change in the substance after it has passed through the membrane. If this change makes it insoluble or charged (to prevent diffusion through the membrane) it is semantics whether this is transport against a concentration gradient or not.

A possible mechanism for iron transport into cells or subcellular organelles is as follows. Fe^{3+} is tightly bound to carrier proteins which enable the ion to penetrate through the membrane. Once inside the compartment the ferric iron is reduced to ferrous and Fe^{2+} is readily dissociated from the carrier. Free carrier returns to collect more Fe^{3+}. The reductive process is the chemical step coupled to transport.

The systems of active transport which are receiving most attention are those which may give information about the sodium pump involved in the uphill movement of Na^+ ion from inside resting nerve fibres to the outside. It is known that ATP is involved in some chemical reaction which is coupled to this process. The physiological background to these phenomena is very elegantly described by Katz (1966). A similar system which is more susceptible to chemical investigations is the exchange between Na^+ and K^+ through the membrane of the red blood cell. A scheme has been proposed in which the free carrier has preferential affinity for K^+ and a phosphorylated carrier binds Na^+. The coupled chemical reaction is involved in the phosphorylation of the carrier inside the cell. Outside the cell there is K^+ stimulated hydrolysis of phosphate linkage and the free carrier binds K^+ and enters the cell (see Katchalsky and Spangler, 1968; Hill, 1968).

CHAPTER 6

Enzyme Kinetics: Steady-State

6.1. RATE EQUATIONS

6.1.1. Mass Law and Order of Reaction

In Section 2.2.1 the law of mass action has been used to explain the relation between reaction rates and equilibria. In a simple isomerization reaction

$$A \underset{k_{-1}}{\overset{k_1}{\rightleftharpoons}} B$$

the equilibria and rates are related by

$$K = \frac{k_1}{k_{-1}} = \frac{\bar{C}_B}{\bar{C}_A}$$

where bars denote equilibrium concentrations. If $k_{-1} \ll k_1$ the reaction will proceed from $C_A = C_A^\circ$ to $C_A = 0$ and $C_B^\circ = 0$ to $C_B = C_A^\circ$, at any time t when the concentration of A is C_A^t the velocity is defined by

$$\frac{dC_A}{dt} = -k_1 C_A^t \quad \text{and} \quad \frac{dC_B}{dt} = k_1 C_A^t \tag{6.1}$$

The order of a reaction is determined by the power of the concentration term in the rate equation. The expression (6.1) is that for a first-order reaction. The units for a first-order rate constant are derived from moles of product per second per mole of reactant:

$$\text{Molarity sec}^{-1} = k \text{ molarity}$$

$$k = \text{sec}^{-1}$$

A reaction of the type

$$A + A \rightarrow A_2$$

or

$$A + B \rightarrow C$$

will be second order if one single reaction step is involved in the process. The rate will be proportional to C_A^2 or $C_A C_B$

$$\frac{dC_{A_2}}{dt} = k C_A^2 \quad \text{or} \quad \frac{dC_C}{dt} = k C_A C_B \tag{6.2}$$

The units of a second-order reaction are derived from

$$\frac{\text{Molarity}}{\text{Seconds}} = k(\text{molarity})^2$$

$$k = (\text{molarity})^{-1}\,\text{sec}^{-1}$$

A reaction of the stoichiometry $A + B \rightarrow C$ can be first order if $A + B$ go first to an intermediate AB and a rate limiting reaction

$$AB \rightarrow C$$

determines the appearance of C. The rate dC_C/dt need then, of course, not be equal to $-dC_A/dt$. Such a process is bimolecular but first order. For unimolecular reactions the terms molecularity and order are equivalent.

Reactions of higher than second order are rare and need not be considered here. However, zero-order reactions are of special importance in the study of enzyme reactions. Their main characteristic is that the rate is constant as the reaction proceeds. The mechanism which accounts for zero order kinetics will be discussed in Section 6.2.4.

6.1.2. Calculation of Rate Constants and Halftimes

The rate constants of reactions or of individual steps of complex chemical processes are important characteristics of the mechanism. The dependence of individual rate constants on many different conditions of the reaction is the principal information for the elucidation of reaction mechanisms by kinetic investigation. The accurate evaluation of the order and magnitude of rate constants requires careful attention.

The way to proceed is to test which rate equation (first or second order, for instance) provides the best fit for the data. Experiments usually provide values for the concentration of the reactant at a given time t. In the ideal case the record of some optical or other physical signal is obtained as a continuous function of time. This physical signal should be directly proportional to the concentration of one of the reactants or products. For a first-order process $(A \rightarrow B)$ the record of appearance of B or disappearance of A gives integrated results of the total amount changed up to the chosen point in time. The differential equations (6.1) and (6.2) could only be used if the rate of change at each point in time is evaluated by drawing a tangent at that point—not a very accurate procedure. The direct use of the integrated values is possible with the integrated rate equation, which is for a first-order process:

$$-\int \frac{dC_A}{C_A} = \int k\,dt$$

$$\ln\frac{C_A^\circ}{C_A} = kt \quad \text{or} \quad \ln C_A^\circ - \ln C_A = kt \tag{6.3}$$

A plot of $\ln C_A$ against t will have the slope k. If values for C_B are obtained as a function of time and C_B is assumed zero at $t = 0$,

$$\ln C_A^\circ - \ln (C_A^\circ - C_B) = kt \quad \text{or}$$

$$\ln C_B^\infty - \ln (C_B^\infty - C_B) = kt$$

where C_B^∞ is the final concentration, which has to be determined if C_A° is not known.

Guggenheim devised an algebraic procedure which makes it possible to obtain rate constants for first-order reactions when neither C_A° nor the end of the reaction are known. This method is illustrated in Figure 21. The Guggenheim procedure can be extended (see p. 202) to the calculation of rate constants for first-order reactions which are followed by a very much slower reaction or a drift.

A distinct property of the first-order rate equation is that it is not dependent on the absolute concentration but on the ratio of the concentration at time t to the final concentration (see equation (6.3)). This also shows up in the fact

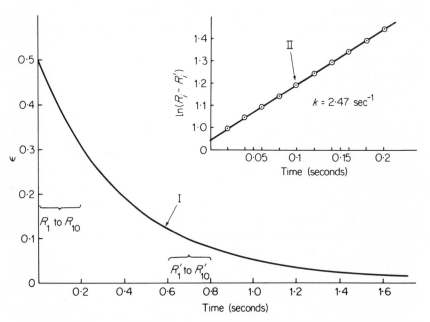

Figure 21   Curve I is a continuous record of the progress of the reaction $A \rightarrow B$; the extinction ε is proportional to the concentration C_A. The Guggenheim method for the calculation of first-order rate constants is illustrated in the plot (II) of $\ln (R_i - R_i')$ against time. The regions over which readings R_i and R_i' are taken are indicated, i's go from 1 to 10

that the first-order rate constant does not have any concentration units, as shown in Section 6.1.1. Consequently it is not necessary to convert the signal measured into concentration units as long as it is linearly proportional to concentration.

One can only have confidence in a reaction being truly first order if the points fit equally well along the straight line of a first-order plot all the way to at least 95 per cent. of the complete reaction.

Another test for first-order behaviour is the determination of successive halftimes. The halftime ($t_{\frac{1}{2}}$) is the time taken for completion of half the remaining reaction from any starting point along the reaction path. From the equation

$$k(t_2 - t_1) = \ln C_A^{t_1} - \ln C_A^{t_2}$$

(t_1 is the chosen starting point when A is at concentration C_A^t) it can be derived for the condition $t_{\frac{1}{2}} = t_2 - t_1$ and $C_A^{t_2} = \frac{1}{2}C_A^{t_1}$ that

$$kt_{\frac{1}{2}} = \ln 2$$

Therefore,

$$t_{\frac{1}{2}} = \frac{\ln 2}{k} = \frac{0.69}{k} \tag{6.4}$$

If halftimes calculated from various starting points along the reaction record remain constant the reaction is likely to be first order.

Second-order reactions are, if at all possible, studied under one of two special conditions. The first of these is called the pseudo first-order condition, which is defined for the reaction

$$A + B \rightarrow C$$

by the condition

$$C_A^\circ \gg C_B^\circ \quad \text{or} \quad C_B^\circ \gg C_A^\circ$$

Taking the case of the first of these conditions, C_A° so large that it can be regarded as a constant throughout the reaction, one can write for the rate equation

$$\ln C_B^\circ - \ln C_B = C_A^\circ kt \tag{6.5}$$

The rate constant obtained from a first-order plot will be $C_A^\circ k$, where k is the second-order rate constant. For a true second-order reaction, measured under these conditions, the slope of the first-order plot must be accurately proportional to C_A° over a fairly wide range of C_A°.

The second condition often used for evaluating second-order rate constant is

$$C_A^\circ = C_B^\circ$$

This condition simplifies the second-order rate equation

$$\frac{dC_C}{dt} = kC_A^2$$

which integrates to give

$$\int_{C_A^\circ}^{C_A^t} \frac{dC_A}{C_A^2} = k \int_0^t dt$$

$$\frac{1}{C_A} - \frac{1}{C_A^\circ} = kt \tag{6.6}$$

A plot of $1/C_A$ against time will have slope k. When the reaction is half complete

$$C_A = \tfrac{1}{2}C_A^\circ$$

and it follows that

$$\frac{2}{C_A} - \frac{1}{C_A^\circ} = kt_{\frac{1}{2}}$$

$$t_{\frac{1}{2}} = \frac{1}{kC_A^\circ} \tag{6.7}$$

This derivation shows that in the case of a second-order reaction the halftime is dependent on the starting concentration and successive halftimes become longer and longer.

There are occasions when it is not possible to get data for a second-order process under either of two special conditions given above. The integration of the general equation gives

$$kt = \frac{1}{C_A^\circ - C_B^\circ} \ln \frac{C_B^\circ(C_A^\circ - C_C)}{C_A^\circ(C_B^\circ - C_C)} \tag{6.8}$$

or

$$kt = \frac{1}{C_A^\circ - C_B^\circ} \ln \frac{C_B^\circ C_A}{C_A^\circ C_B} \tag{6.9}$$

6.2. COMPLEX REACTIONS

6.2.1. Reversible Reactions

If the simplifying assumption $k_{-1} = 0$, made in the above derivations, is removed for the reaction $A \rightleftharpoons B$ it follows that at $t = \infty$

$$K = \frac{\bar{C}_B}{\bar{C}_A} = \frac{k_1}{k_{-1}} \tag{6.10}$$

with $\bar{C}_B$ and $\bar{C}_A$ the equilibrium concentrations of the reactants. At equilibrium there is no net change in concentration since $k_1\bar{C}_A = k_{-1}\bar{C}_B$. The rate of approach to equilibrium can be derived from these considerations as follows:

$$k_1(C_A^\circ - \bar{C}_B) - k_{-1}\bar{C}_B = 0$$

$$C_A^\circ = \frac{k_{-1}}{k_1}\bar{C}_B + \bar{C}_B$$

$$\frac{dC_B}{dt} = k_1(C_A^\circ - C_B) - k_{-1}C_B$$

$$\frac{dC_B}{dt} = k_{-1}\bar{C}_B + k_1\bar{C}_B - k_1C_B - k_{-1}C_B$$

$$\frac{dC_B}{dt} = (k_{-1} + k_1)(\bar{C}_B - C_B) \qquad (6.11)$$

This equation has the form of a first-order equation. The reaction from C_B to $\bar{C}_B$ proceeds with an apparent first-order rate constant $(k_{-1} + k_1)$. When k_{-1} is taken as zero the equation is equivalent to (6.1). The rate constants can be evaluated from $k_1 + k_{-1}$ obtained from a first-order plot of $\ln(\bar{C}_B - C_B)$ against t and the equilibrium constant (see equation (6.10)).

The rate equations become much more complex when second-order steps are involved in reversible reactions. Various processes and methods for solving the equations describing them are discussed by Frost and Pearson (1961). One case is of particular importance for the study of enzyme systems:

$$E + A \underset{k_{-1}}{\overset{k_1}{\rightleftharpoons}} EA \qquad (6.12)$$

where E stands for enzyme and A can be a substrate, inhibitor or activator. The complex EA is formed in a second-order reaction and dissociates to free enzyme and A in a first-order process. In principle it is possible to evaluate k_1 and k_{-1} by making use of the pseudo first-order conditions

$$C_A \gg C_E \quad \text{and} \quad k_1C_A \gg k_{-1} \qquad (6.13)$$

for the determination of k_1 (see Section 6.1.2). If the equilibrium constant for the process

$$K = \frac{C_{AE}}{C_E C_A} = \frac{k_1}{k_{-1}}$$

is known k_{-1} can be calculated.

In practice it is often not possible to fulfil both the conditions given in (6.13) because the reaction may become too rapid to be measured (however, see methods described in Chapter 8). If the reaction is measured under conditions $C_A^\circ = C_E^\circ$ and at absolute concentrations at which equilibrium is reached with all components at measurable concentrations, more complex algebraic expressions have to be used to evaluate the rate constants. For the reaction given in scheme (6.12) the integrated rate equation is

$$\ln \frac{\bar{C}_{EA}(C_E^2 - C_{EA}\bar{C}_{EA})}{(\bar{C}_{EA} - C_{EA})C_E^2} = k_1 \frac{C_E^2 - \bar{C}_{EA}^2}{\bar{C}_{EA}} t \tag{6.14}$$

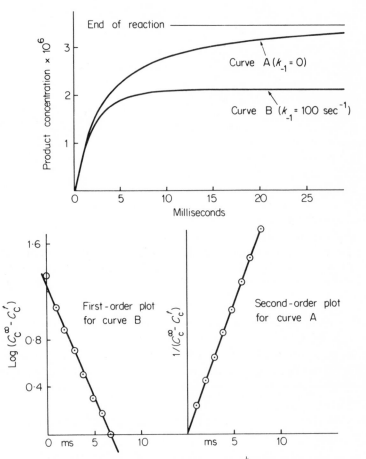

Figure 22   The progress of the reaction $A + B \overset{k_1}{\underset{}{\rightleftharpoons}} C$ was simulated on an analog computer ($C_A^\circ = C_B^\circ = 3.5 \times 10^{-6}$ M, $k_1 = 10^8$ M^{-1} sec^{-1}). For curve A $k_{-1} = 0$ and for curve B $k_{-1} = 100$ sec^{-1}. Curve A can be plotted as a second-order reaction. Curve B, however, gives an apparently quite good first-order plot, though careful examination reveals systematic deviation from this interpretation

To evaluate k_1 from measurements of the change in C_{EA} or C_E with time one also needs to know the equilibrium constant to calculate $\bar{C}_{EA}$ from the known values for $C_E^\circ = C_A^\circ$. A practical point is demonstrated in Figure 22. Even under conditions of equilibrium constant and concentrations resulting in the reaction proceeding to 62 per cent. completion to form EA the process could be erroneously interpreted as a first-order reaction. The determination of the order of a reaction requires a detailed study of the concentration dependence of all reactants to avoid misinterpretations. This is of special importance for steps of unknown total amplitude.

6.2.2. Consecutive Reactions

Consecutive reactions leading to steady-state levels of intermediates have been discussed in Section 5.1.3. In the present section a consecutive reaction with a transient, instead of a steady-state intermediate, will be analysed and compared with the steady-state approximation.

The time course of the concentration changes of the three reactants of the process

$$A \xrightarrow{k_1} B \xrightarrow{k_2} C$$

is described by the following equations

$$-\frac{dC_A}{dt} = k_1 C_A$$

$$C_A = C_A^\circ e^{-k_1 t} \tag{6.15}$$

$$\frac{dC_B}{dt} = k_1 C_A - k_2 C_B = k_1 C_A^\circ e^{-k_1 t} - k_2 C_B$$

$$C_B = k_1 C_A^\circ \frac{e^{-k_1 t} - e^{-k_2 t}}{k_2 - k_1} \tag{6.16}$$

for the condition $C_B^\circ = 0$

$$C_C = C_A^\circ - C_A - C_B$$

$$C_C = C_A^\circ \left[1 + \frac{1}{k_1 - k_2}(k_2 e^{-k_1 t} - k_1 e^{-k_2 t}) \right] \tag{6.17}$$

The computer representation of the reaction discussed above illustrates a number of salient points (see Figure 23). The disappearance of A follows a typical first-order process. The initial slope of the appearance of B can also be described by the same first-order rate constant k_1. However, if the maximum of the curve representing C_B were taken as an end-point for the calculation of a rate constant a first-order k_{app} would be obtained which is larger than the

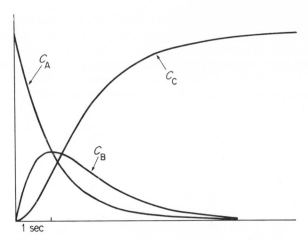

Figure 23 Analog Computer simulation of the concentration changes during the reaction $A \xrightarrow{k_1} B \xrightarrow{k_2} C$. Both steps are taken as irreversible and $k_1 = k_2 = 1 \sec^{-1}$.

first-order constant in the equation

$$\frac{dC_B}{dt} = k_1 C_A$$

C_B is at a maximum when

$$k_1 C_A = k_2 C_B$$

For the case illustrated in Figure 23 $k_1 = k_2$ and at the inflection point for C_B the concentrations of C_B and C_A are equal. If the disappearance of A and the appearance of B can both be followed, k_1 and k_2 can be calculated without recourse to equation (6.17). The complex equation is needed when the observed parameters are the concentration changes in A and C.

The reaction discussed above can be used to illustrate the steady-state approximation for some kinetic investigations of reaction mechanisms. For the particular case $k_1 = k_2$ the steady-state in C_B is momentarily reached when $dC_B/dt = 0$. If one considers a frequently observed modification of the scheme $A \rightarrow B \rightarrow C$ by including a reversible step with $(k_1 + k_{-1}) > k_2$

$$A \underset{}{\overset{k_1}{\rightleftharpoons}} B \xrightarrow{k_2} C$$

one finds two interesting new features. If the disappearance of A is observed the time course no longer corresponds to a single first-order reaction, instead two phases correspond to the coupled relaxations of the two steps. This point will be discussed in some detail in Chapter 8.

The second feature of sequential equilibria is that under many conditions C_B is essentially constant for a sufficient part of the total reaction that it can be considered to be in a steady state. For the simple irreversible consecutive reaction considered first, the steady-state approximation gives

$$k_1 C_A - k_2 C_B = 0$$

$$\frac{C_B}{C_A} = \frac{k_1}{k_2}$$

$$C_A = C_A^\circ e^{-k_1 t}$$

$$C_B = C_A^\circ \frac{k_1}{k_2} e^{-k_1 t}$$

$$C_C = C_A^\circ - C_A - C_B = C_A^\circ \left[1 - e^{-k_1 t}\left(1 + \frac{k_1}{k_2}\right)\right] \quad (6.18)$$

These equations for the concentrations of C_B and C_C become equivalent to the exact solution when t is large compared with $1/k_2$.

For further discussions of series reactions Frost and Pearson (1961) and Chapter 8 should be consulted.

6.2.3. Parallel Reactions

Parallel reactions can have a common substrate

$$A \rightarrow B$$
$$A \rightarrow C$$

or

$$A + X \rightarrow B$$
$$A + Y \rightarrow C$$

Alternatively, one can have parallel reactions with a common product

$$X \rightarrow A$$
$$Y \rightarrow A$$

or

$$A + X \rightarrow B$$
$$A + Y \rightarrow B$$

The rate equations for all these cases are easy to derive by methods already discussed in the preceding sections of this chapter.

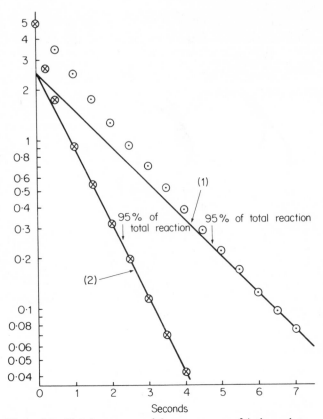

Figure 24   The time course of the appearance of A through two parallel and simultaneous reactions $X \xrightarrow{k_1} A$ and $Y \xrightarrow{k_2} A$ was simulated on an analog computer for two sets of rate constant (1) $k_1 = 1 \text{ sec}^{-1}$, $k_2 = 0.5 \text{ sec}^{-1}$ and (2) $k_1 = 5 \text{ sec}^{-1}$, $k_2 = 1 \text{ sec}^{-1}$. In both cases $C_X^\circ = 2.5$, $C_Y^\circ = 2.5$ and $C_A^\infty = 5$, all in arbitrary units. Plots of $\log(C_A^\infty - C_A^t)$ against time are given by $\odot$ for condition (1) and by $\otimes$ for condition (2). With such systems of two first order processes it is, in principle, possible to calculate both rate constants and the contribution of C_X° and C_Y° to the total production of A (i.e. to C_A^∞). The procedure is illustrated by plot (2) where the last third of the total reaction follows a simple first-order course with the solid line (which corresponds to a rate constant of 1 sec^{-1}) extrapolating to the contribution of C_Y°. The values for the contribution of C_Y are then subtracted from the time course of the whole reaction and k_1 is calculated from the log plot of the residue.

For the points of plot (1) this procedure is only possible if *very* accurate data are obtained for the last few per cent. of the reaction. In the above diagram the theoretical line is drawn for
$$k_2 = 0.5 \text{ sec}^{-1} \text{ to the known concentration of } C_Y^\circ$$

The main problem with the analysis of such systems lies in the resolution of the individual rate constants from the observed records. As an example the computer representation of two parallel first-order reactions with a common product, will be considered. The detailed data given in Figure 24 are intended to illustrate how difficult it is to resolve two superimposed exponentials if their time constants differ by less than a factor of three.

6.2.4. Catalysis

Most of what follows in this book is concerned more or less directly with the kinetic behaviour of enzyme systems and catalysis by individual enzymes. This section is only concerned with a few general kinetic consequences of catalysis.

One definition of a catalyst is: a reagent which takes part in a reaction while itself remaining at a constant concentration, which is small compared with that of the other reactants. Another definition is: a catalyst accelerates the forward and reverse reaction by the same factor and therefore does not affect the final equilibrium. In the discussion of enzyme reactions a distinction will be made between conditions when the enzyme is present at catalytic (very low) concentrations and other special cases when stoichiometric reactions of enzymes are studied at high concentrations. The definition of enzymes as catalysts only applies to the first of these conditions.

If C is the catalyst accelerating the interconversion

$$A \rightleftharpoons B$$

some transient contacts of the type AC and BC must be made. If one writes the

$$A + C \overset{k_1}{\rightleftharpoons} AC \overset{k_2}{\rightleftharpoons} BC \overset{k_3}{\rightleftharpoons} B + C$$

$$K_1 = \frac{k_1}{k_{-1}}, \qquad K_2 = \frac{k_2}{k_{-2}}, \qquad K_3 = \frac{k_3}{k_{-3}}$$

Then the equilibrium for the overall reaction K is given by

$$K = \frac{C_B}{C_A} = K_1 K_2 K_3 = \frac{k_1 k_2 k_3}{k_{-1} k_{-2} k_{-3}} \tag{6.19}$$

This relation is similar to the Haldane relation for enzyme reactions derived in Section 6.3.5. Other aspects of catalysis and its kinetic consequences are discussed in some detail in this and the following chapter. It just remains in this section to point out that catalysis can lead to zero order kinetics

$$\frac{dC_B}{dt} = \text{constant}$$

This relation will apply for a catalytic system if the substrate concentration C_A is large enough to be a source in the sense defined in Section 5.1.3 and some step after formation of AC is rate limiting. The zero order constant will be a composite of a first-order rate constant and the enzyme concentration and will have the units M sec^{-1}.

6.3. FROM MICHAELIS TO STEADY-STATE KINETICS

6.3.1. The Michaelis Equation

The early history of enzyme kinetics is admirably reviewed by Segal (1959). For an understanding of the principles involved one does not need to go further back than the treatment of Michaelis and Menten. A discussion of the reasoning behind Michaelis and Menten's derivations should be helpful in the definition of some of the basic assumptions of enzyme kinetics.

From the phenomenon illustrated in Figure 25 it was deduced that enzyme-catalysed conversion of substrate into product proceeds through the formation of an intermediate: the enzyme–substrate complex. Figure 25 shows that as substrate concentration C_S is increased the velocity of the appearance of product reaches a maximum. Only at very low substrate concentrations is there a linear first-order dependence of dC_P/dt upon C_S.

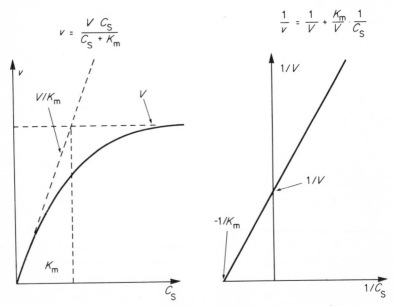

Figure 25   Plots of direct and reciprocal relations between substrate concentration, C_S, initial velocities, v, maximum velocity, V, and Michaelis constant, K_m

Schematically this is expressed as follows:

$$E + S \underset{}{\overset{k_1}{\rightleftharpoons}} ES \overset{k_2}{\rightarrow} E + P$$

The total enzyme active site concentration C_E° is the sum

$$C_E^\circ = C_E + C_{ES}$$

where C_E stands for concentration of free enzyme and C_{ES} for concentration of enzyme–substrate complex. The following two assumptions were made by Michaelis and Menten:

(1) In the steady state the enzyme exists essentially only in two forms E and ES.

(2) ES is in true equilibrium with E + S during the time period of the observation.

Both these assumptions depend on k_2, the rate constant for the decomposition of the first complex between enzyme and substrate, being very much smaller than any of the other rate constants. As a first approximation the reverse reaction of enzyme recombining with product is neglected because C_P is very small during the time course of the measurement. From these assumptions it follows that the velocity

$$v = \frac{dC_P}{dt} = k_2 C_{ES} \tag{6.20}$$

and the maximum velocity V which is attained at high substrate concentration when

$$C_{ES} = C_E^\circ \qquad V = k_2 C_E^\circ \tag{6.21}$$

The equilibrium between enzyme–substrate complex and free E and S is described by the Michaelis constant K_m

$$K_m = \frac{(C_E^\circ - C_{ES})C_S}{C_{ES}} \tag{6.22}$$

C_S is regarded as sufficiently large to be a constant at any given substrate concentration. It follows from equations (6.20), (6.21) and (6.22) that

$$K_m C_{ES} = C_S C_E^\circ - C_{ES} C_S$$

$$K_m \frac{v}{k_2} = C_S \frac{V}{k_2} - C_S \frac{v}{k_2}$$

$$v = \frac{V C_S}{K_m + C_S} \tag{6.23}$$

It will be seen that an equation of this form describes much more complex enzyme reactions, but the physical significance of the two constants (Michaelis parameters) changes as the assumptions, introduced above, are removed.

6.3.2. The Determination of the Michaelis Parameters

Equation (6.23) relates velocity to substrate concentration through the two Michaelis parameters characteristic for an enzyme under a particular set of conditions (temperature, pH, ionic composition). It is readily seen that when

$$C_S \ll K_m \qquad v = C_S \frac{V}{K_m} \text{ (first-order conditions)}$$

$$C_S \gg K_m \qquad v = V$$

$$C_S = K_m \qquad v = V/2$$

Equation (6.23) can be rewritten in several ways to give linear plots of experimental data. Three examples are given in equations (6.24), (6.25) and (6.26).

$$\frac{1}{v} = \frac{K_m}{V} \frac{1}{C_S} + \frac{1}{V} \tag{6.24}$$

If $1/v$ (ordinate) is plotted against $1/C_S$ (abscissa), the ordinate intercept is $1/V$ and the slope is K_m/V.

$$\frac{C_S}{v} = \frac{K_m}{V} + \frac{C_S}{V} \tag{6.25}$$

If C_S/v (ordinate) is plotted against C_S (abscissa), the ordinate intercept is K_m/V and the slope $1/V$.

$$v = V - \frac{v}{C_S} K_m \tag{6.26}$$

If v (ordinate) is plotted against v/C_S (abscissa), the ordinate intercept is V and the slope is K_m. Of these three graphical methods for evaluating the Michaelis parameters the one described by equation (6.24), the Lineweaver–Burk plot, is the most widely used and the most unsatisfactory. Merit judgements on the three methods are based on a number of desirable features such as V and K_m in the numerator, C_S in the numerator and the dependent variable only on one side of the equation. The second reciprocal equation (6.25) is the only one which has two of the desired properties and is preferred by some investigators. However, the advantage of the third equation (6.26) of not using reciprocals of V and K_m is on its own sufficiently great to warrant the use of that method.

The accurate determination of the Michaelis parameters depends not only, and perhaps less so, on the best equation but even more on the provision of the best possible data. The range of substrate concentrations used for velocity measurements, as well as the determination of the velocity under conditions corresponding as precisely as possible to the constant substrate concentration, needs some further discussion. Taking the second point first, this can be illustrated with a simple calculation. This calculation requires that an approximate value for K_m is estimated. Inspection of the tangents, from which the initial velocities are calculated for each C_S, permits calculation of the amount of substrate ΔC_S used up during the interval Δt (see Figure 26).

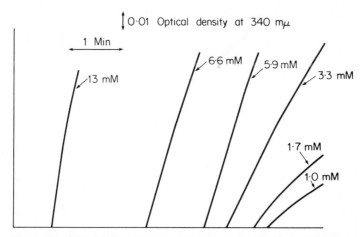

Figure 26   This family of curves represents records of the oxidation of NADH by lactate dehydrogenase (M_4) at pH 7 at different initial pyruvate concentrations, as indicated on each line. Clearly, initial rates are progessively more difficult to determine as the substrate concentration is lowered. Systematic errors are likely to arise from the fact that estimates of the initial rates will be progessively lower than the true value as the substrate concentration is decreased

One can then check whether linearity of the initial slope is satisfied by the relation

$$\frac{C_S V}{K_m + C_S} \quad \text{is within 1 per cent. equal to} \quad \frac{(C_S - \Delta C_S)V}{K_m + (C_S - \Delta C_S)}$$

The effect of product formed has also got to be taken into consideration. If K_m is very small and the sensitivity for recording ΔC_S not very great, one may have to use a reiterative procedure for the evaluation of correct Michaelis parameters.

The range of initial substrate concentrations used for the determination of initial velocities should be evenly spread around K_m from $C_S = 5K_m$ to $C_S = K_m/5$.

6.3.3. The Integrated Rate Equation

Preliminary estimates of the Michaelis parameters can be obtained in a variety of ways. If neither excess substrate nor product produce significant inhibition of the enzyme under investigation, the record of a single complete reaction should give approximate values for K_m and V. Simple inspection of the substrate concentration at half maximum velocity gives a useful value for K_m if one is sure that maximum velocity was attained at the beginning of the reaction. One can be satisfied about this if truly zero order rates are obtained at the beginning. A more general procedure is the use of the integrated rate equation for the interpretation of a single complete reaction. Equation (6.23) can be written for $C_S^\circ \gg C_E^\circ$, where C_S° is the initial substrate concentration and C_P^t, the product concentration at time t

$$v = \frac{V(C_S^\circ - C_P^t)}{K_m + C_S^\circ - C_P^t}$$

this can be integrated with respect to time to give

$$Vt = C_P^t + K_m \ln \frac{C_S^\circ}{C_S^\circ - C_P^t} \quad \text{or} \quad \left(\ln \frac{C_S^\circ}{C_S^\circ - C_P^t} \right) \Big/ t = \frac{V}{K_m} - \frac{1}{K_m} \frac{C_P^t}{t} \quad (6.27)$$

A plot of $\left(\ln \dfrac{C_S^\circ}{C_S^\circ - C_P^t} \right) \Big/ t$ against C_P^t/t will have a slope of $-1/K_m$ and an intercept of V/K_m. If product inhibition or appreciable reversible reactions occur the plot will not remain linear for the whole course of the reaction.

6.3.4. The Steady-state Approximation

Briggs and Haldane (1925) have shown that it is unnecessary to make the assumption that enzyme and substrate are in thermodynamic equilibrium with the enzyme–substrate complex to derive an equation of the form of the Michaelis equation. They state that some time after enzyme and substrate are mixed the concentrations of free enzyme sites and of substrate complexed sites will reach a steady state. As discussed in Section 5.1.3, the maintenance of a steady state of intermediates over a finite period of observation depends on the substrate pool being large enough to be essentially constant. The steady state is defined by equations based on the condition that all intermediate concentrations are constant:

$$dC_{ES}/dt = 0 = k_1 C_E C_S - (k_2 + k_{-1})C_{ES} \quad (6.28)$$

$$dC_E/dt = 0 = (k_2 + k_{-1})C_{ES} - k_1 C_E C_S \quad (6.29)$$

To these equations of the steady-state conditions can be added the equation of conservation of enzyme sites

$$C_E^\circ = C_{ES} + C_E \qquad (6.30)$$

The equation for K_m can be given the physical meaning of a steady-state constant if steady-state concentrations instead of equilibrium concentrations are used in

$$K_m = \frac{(C_E^\circ - C_{ES})C_S}{C_{ES}} \qquad (6.22)$$

From equations (6.22), (6.28), (6.29) and (6.30) it follows that

$$K_m = \frac{k_2 + k_{-1}}{k_1} \qquad (6.31)$$

The difference between the steady-state K_m and the equilibrium constant k_{-1}/k_1 disappears when $k_2 \ll k_{-1}$. Substituting v for $k_2 C_{ES}$ and solving equations (6.28), (6.29) and (6.30) for the velocity of product formation per mole of enzyme sites results in the following relation

$$\frac{v}{C_E^\circ} = \frac{k_2 C_S}{[(k_2 + k_{-1})/k_1] + C_S} \qquad (6.32)$$

A comparison of equation (6.32) with the Michaelis equation (6.23) shows that

$$k_2 C_E^\circ = V \quad \text{and} \quad (k_2 + k_{-1})/k_1 = K_m$$

It now remains to remove the second assumption of the Michaelis–Menten theory in its original form, namely that the enzyme only exists in two forms: free enzyme and one enzyme–substrate complex. As will be seen from evidence presented in Chapter 8, the enzyme is usually distributed among several forms of enzyme–substrate and enzyme–product complexes as well as free enzyme. The simplest scheme which can be used to demonstrate all the algebraic consequences of more realistic models is one which includes an enzyme–product complex as well as an enzyme–substrate complex:

$$E + S \underset{}{\overset{k_1}{\rightleftharpoons}} ES \underset{}{\overset{k_2}{\rightleftharpoons}} EP \underset{}{\overset{k_3}{\rightleftharpoons}} E + P$$

The three equations defining the steady-state conditions are

$$dC_E/dt = k_3 C_{EP} + k_{-1} C_{ES} - C_E(k_1 C_S + k_{-3} C_P) = 0 \qquad (6.33)$$

$$dC_{ES}/dt = k_1 C_E C_S + k_{-2} C_{EP} - (k_{-1} + k_2)C_{ES} = 0 \qquad (6.34)$$

$$dC_{EP}/dt = k_{-3} C_E C_P + k_2 C_{ES} - (k_{-2} + k_3)C_{EP} = 0 \qquad (6.35)$$

If the molarity of enzyme sites $C_E^\circ = C_E + C_{ES} + C_{EP}$ is considered constant and this relation is used to eliminate C_E from the steady-state equations one can proceed towards obtaining an expression for v/C_E° in terms of rate constants and C_S only. The rate equation is first derived for the case in which C_P is essentially zero during the course of the observation and all terms containing C_P are neglected. The velocity of formation of product is

$$v = dC_P/dt = k_3 C_{EP} = k_2 C_{ES} - k_{-2} C_{EP} \tag{6.36}$$

This merely restates the steady-state definition that the rates of formation and decomposition of each intermediate which provides the flux of reactants through the system are equal. From equations (6.33) to (6.36) one can derive

$$\frac{v}{C_E^\circ} = \frac{k_1 k_2 k_3 C_S}{k_{-1}k_{-2} + k_{-1}k_3 + k_2 k_3 + k_1(k_2 + k_{-2} + k_3)C_S} \tag{6.37}$$

which can be rearranged to give

$$\frac{v}{C_E^\circ} = \frac{\dfrac{k_2 k_3}{k_2 + k_{-2} + k_3} C_S}{\dfrac{k_{-1}k_{-2} + k_{-1}k_3 + k_2 k_3}{k_1(k_2 + k_{-2} + k_3)} + C_S} \tag{6.38}$$

Equation (6.38) is of the general form of (6.23) with

$$K_m = \frac{k_{-1}k_{-2} + k_{-1}k_3 + k_2 k_3}{k_1(k_2 + k_{-2} + k_3)} \tag{6.39}$$

and

$$V = C_E^\circ \frac{k_2 k_3}{k_2 + k_{-2} + k_3} \tag{6.40}$$

However complex the reaction mechanism, as long as it can be described by a velocity equation of this form one can evaluate K_m and V from rate measurements by the methods of Section 6.3.2. The physical reason for the complex expression for K_m, for the case of reactions with several intermediates, is that the steady-state concentrations of the intermediates depend on all the rate constants. The steady-state concentrations of the intermediates at a particular concentration C_S in turn determine the reaction velocity.

Only in special cases (see Section 6.3.1) is the Michaelis constant equal to $K_S = k_{-1}/k_1$ the substrate dissociation constant. The distinction between enzyme–substrate dissociation constant and K_m has become more difficult since substrate binding too can now be shown to be a complex sequence of events (see Section 8.3.1). In the relatively simple case, with only one step involved in enzyme–substrate combination, one can demonstrate how K_m

can change through change in substrate reactivity by changing k_2 in a reaction with k_{-2} and k_{-3} sufficiently slow to be neglected

$$E + S \underset{}{\overset{k_1}{\rightleftharpoons}} ES \overset{k_2}{\rightarrow} EP \overset{k_3}{\rightarrow} E + P$$

K_m and V of equation (6.38) simplify under these conditions to

$$V = C_E^\circ \frac{k_2 k_3}{k_2 + k_3}$$

$$K_m = \frac{k_{-1} + k_2}{k_1} \frac{k_3}{k_2 + k_3} \tag{6.41}$$

If all rate constants are kept constant except k_2, which characterizes the rate of the chemical transformation $ES \rightarrow EP$, the following changes occur in K_m:

(1) $k_2 \ll k_{-1}$ and $k_2 \ll k_3$

$$K_m = \frac{k_{-1}}{k_1} = K_S$$

(2) k_2 larger, equal to or smaller than k_{-1} and $k_2 \ll k_3$

$$K_m = \frac{k_{-1} + k_2}{k_1} > K_S$$

(3) k_2 larger, equal to or smaller than k_{-1} and $k_2 \geqslant k_3$

$$K_m = \frac{k_{-1} + k_2}{k_1} \frac{k_3}{k_2 + k_3} < K_S$$

In the last case K_m becomes progressively smaller than K_S as k_2 increases.

While the significance of K_m for the elucidation of enzyme mechanisms will become smaller as methods for the elucidation of individual steps of the reaction become available, the Michaelis constant is an important parameter describing the physiological behaviour of an enzyme. The range of substrate concentration in the region of K_m is the most suitable level for substrate control of enzyme activity. This point as well as the discussion of deviations from Michaelis-type kinetics, will be taken up again in Section 6.6.2.

6.3.5. The Haldane Relation for Reversible Reactions

Haldane (1930) has applied the steady-state approximation to single substrate reactions with all steps considered as essentially reversible:

$$E + S \overset{k_1}{\rightleftharpoons} ES \overset{k_2}{\rightleftharpoons} EP \overset{k_3}{\rightleftharpoons} E + P$$

For this scheme we can derive V^S and K_m^S for substrate in the absence of product

$$V^{S} = C_{E}^{\circ} \frac{k_2 k_3}{k_2 + k_{-2} + k_3} \qquad K_{m}^{S} = \frac{k_{-1}k_{-2} + k_{-1}k_3 + k_2 k_3}{k_1(k_2 + k_{-2} + k_3)}$$

For the reverse reaction V^{P} and K_{m}^{P} (the Michaelis parameters for the reaction $P \rightarrow S$) in the absence of substrate the same derivation gives

$$V^{P} = C_{E}^{\circ} \frac{k_{-1}k_{-2}}{k_{-1} + k_2 + k_{-2}} \qquad K_{m}^{P} = \frac{k_{-1}k_{-2} + k_{-1}k_3 + k_2 k_3}{k_{-3}(k_{-1} + k_2 + k_{-2})}$$

For the reaction at finite concentrations of both substrate and product the rate equation is

$$v = k_2 C_{ES} - k_{-2} C_{EP} =$$

$$C_{E}^{\circ} \frac{k_1 k_2 k_3 C_S - k_{-1}k_{-2}k_{-3}C_P}{k_{-1}k_{-2} + k_{-1}k_3 + k_2 k_3 + C_S k_1(k_2 + k_{-2} + k_3) + C_P k_{-3}(k_{-1} + k_2 + k_{-2})}$$

On substituting the Michaelis parameters for the substrate reaction in the absence of product and the Michaelis parameters for the product reaction in the absence of substrate, the following expression is obtained

$$v = \frac{V^{S}K_{m}^{P}C_S - V^{P}K_{m}^{S}C_P}{K_{m}^{S}K_{m}^{P} + K_{m}^{P}C_S + K_{m}^{S}C_P} \qquad (6.42)$$

At equilibrium $v = 0$ and the enumerator of equation (6.42) is zero:

$$V^{S}K_{m}^{P}C_S = V^{P}K_{m}^{S}C_P$$

and the equilibrium constant

$$K = \frac{C_P}{C_S} = \frac{V^{S}K_{m}^{P}}{V^{P}K_{m}^{S}} \qquad (6.43)$$

This is an important result because it helps to cross-check experimental determinations of equilibrium constants and Michaelis parameters. When the expressions for V^{S}, V^{P}, K_{m}^{S} and K_{m}^{P} in terms of the six rate constants are substituted in equation (6.43), we obtain equation (6.19) derived in Section 6.2.4.

6.4. TWO SUBSTRATE SYSTEMS

6.4.1. Introduction to Two Variable Substrates

There are in fact very few truly single substrate enzyme reactions. Hydrolytic reactions, studied in such detail by enzymologists, involve water as a second substrate. Many attempts at discovering a water binding site on enzymes have failed because it is impossible to change the concentration or the properties of water without changing the structure of the enzyme itself. In most investigations of reactions involving water as a substrate it is regarded as at constant concentration and not involved in the kinetic equations. In the hydration reactions catalysed by fumarase, carbonic anhydrase and aconitase the role of water in the mechanism has been largely neglected for the same reason. In the case of the aconitase reaction

$$\text{Citrate} \rightleftharpoons \underset{+\,H_2O}{\text{cis-Aconitate}} \rightleftharpoons \text{Isocitrate}$$

there is some evidence that the water removed from citrate to form cis-aconitate can stay on the enzyme molecule long enough to be reintroduced into cis-aconitate to form isocitrate.

The proton too is a substrate for many enzyme reactions, but its concentration cannot be varied without seriously affecting other parameters. In Section 2.2.5 some examples of the use of indicators for observing the fate of protons were discussed. With most other substrates very marked changes in concentration can also have secondary effects due to changes in ionic strength or the introduction of specific ionic interactions. Such problems are usually not serious as long as one is aware of them.

For most two-substrate reactions it is possible to keep one substrate in such excess that its concentration can be regarded both constant and saturating. This simplifies the kinetics and restricts at the same time the amount of information which can be obtained. It is, however, important to obtain, first of all, data for the initial velocities under the following conditions: for each substrate K_m and the maximum velocity should be determined at saturating concentrations of the other. When K_m for substrate A (K_m^A) is determined below saturating concentration of substrate B (and vice versa for K_m^B) the dependence of K_m for A on the concentration of B contains interesting information about the mechanism. This is one of the main pursuits of steady-state kinetic investigations.

6.4.2. Classification of Mechanisms

The general mechanisms for the reaction of an enzyme with two substrates are first of all divided into sequential:

or substituted enzyme mechanisms:

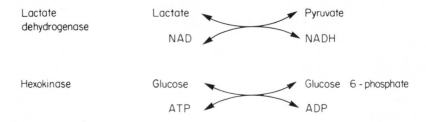

$$E \xrightleftharpoons[AX]{} EAX \xrightleftharpoons[A]{} EX \xrightleftharpoons[B]{} EBX \xrightleftharpoons[BX]{} E$$

Typical sequential mechanisms are those involved in the reactions catalysed by the NAD(P) linked dehydrogenases or the ATP-linked kinases. In these reactions hydrogen and phosphate, respectively, are transferred directly between substrates and nucleotides. Examples of these reactions are

| Lactate dehydrogenase | Lactate ⤢ Pyruvate |
| | NAD ⤢ NADH |

| Hexokinase | Glucose ⤢ Glucose 6-phosphate |
| | ATP ⤢ ADP |

An example of a substituted enzyme reaction is the transaminase:

$$\text{Glutamate} + \text{Enzyme} \rightleftharpoons \text{Enzyme NH}_2 + \alpha\text{-Oxoglutarate}$$

$$\text{Enzyme NH}_2 + \text{Oxaloacetate} \rightleftharpoons \text{Enzyme} + \text{Aspartate}$$

If the method for the derivation of rate equations described in 6.3.4 is applied to a completely general mechanism with all possible pathways operating with defined rate constants an excessive number of simultaneous equations have to be solved. Rate equations for a number of restricted pathways through this maze have been solved.

Sequential mechanisms are divided into ordered and random pathways. In the ordered pathway one substrate has to combine with the enzyme before the other in an obligatory sequence to form the central complex. As a corollary a particular product has to dissociate from the central complex before the other. The treatment of the random pathway mechanism is usually restricted to the case of a rapid equilibrium between both substrate (or products) and enzyme to form the central complex—the interconversion of the two central complexes is rate-limiting. These simplified models have turned out to be very useful but it has to be emphasized that they are only approximations. There is little doubt that in most cases alternative pathways operate, albeit to a limited extent. How good the approximations are depends on the accuracy of the experimental methods used to test them. There are three methods available for the investigation of these mechanisms: first the steady-state analysis discussed in Section 6.4.3, secondly the isotope exchange technique (6.4.4) and thirdly the analysis of transients discussed in Sections 8.2.1 to 8.2.3.

6.4.3. Rate Equations for Two Substrate Mechanisms

The Michaelis constant K_m^A is defined as the concentration C_A which gives, at saturating concentrations of all other substrates, the initial velocity $v = V/2$. As pointed out in Section 6.3.4, for a single substrate the increasing complexity of a mechanism results in more and more constants used to describe K_m and V. Alberty (1953) used the equation

$$v = \frac{V C_A C_B}{K_{AB} + K_m^A C_B + K_m^B C_A + C_A C_B} \tag{6.44}$$

to describe the rates of a two substrate sequential mechanism in terms of the concentrations of the two substrates, the two K_m^S and K_{AB} as an additional constant. It is evident that when $C_B \gg K_{AB}$, K_m^B equation (6.44) reduces to the simple form

$$v = V C_A / [K_m^A + C_A]$$

Another equation, which can be used to express the initial velocities of a sequential two substrate reaction, is that developed by Dalziel (1957):

$$v = C_E^\circ C_A C_B / [C_A C_B \phi_0 + C_B \phi_1 + C_A \phi_2 + \phi_{12}] \tag{6.45}$$

If C_E° is the molarity of enzyme active sites the equation can be normalized to unit enzyme concentration:

$$\frac{v}{C_E^\circ} = 1/[\phi_0 + \phi_1/C_A + \phi_2/C_B + \phi_{12}/C_A C_B] \tag{6.46}$$

When both C_A and C_B are at saturating level,

$$\frac{v}{C_E^\circ} = \frac{V}{C_E^\circ} = \frac{1}{\phi_0} \quad \text{and} \quad \frac{1}{\phi_0} \text{ is the turnover number.}$$

The four Dalziel coefficients can be evaluated even if it is not possible or convenient to study the reaction at saturating substrate concentrations. If a series of initial rate measurements at different C_A and constant C_B are carried out and C_E°/v is plotted against $1/C_A$ a straight line should be obtained. Both the slope and the intercept should be a linear function of C_B. If several such series of measurements are carried out for different constant C_B, the slopes and intercepts can be plotted against $1/C_B$ to give the four coefficients:

Secondary plot of intercepts	slope = ϕ_2	intercept = ϕ_0
Secondary plot of slopes	slope = ϕ_{12}	intercept = ϕ_1

Recently Dalziel (1969) described the extension of this method to the evaluation of the eight coefficients needed to describe the rate of a three substrate reaction. For reversible reactions ϕ'_0, ϕ'_1, etc., can be evaluated in an analogous manner by varying C_X and C_Y in the absence of A and B.

The Alberty equation can be used in a similar way. If K_m^A is determined at a number of different fixed concentrations C_B by the Lineweaver and Burk procedure (6.3.2) a family of straight intersecting lines is obtained in the case of a sequential ordered mechanism (with A first combining with the enzyme) or a random order rapid equilibrium mechanism. The equilibrium binding constant for $A(K_S^A)$ is found by drawing a vertical line from the intersection of the reciprocal plot to the x-axis. At that point $-1/C_A = K_S^A$. In the case of a substituted enzyme mechanism, where only one substrate binds at a time, the Lineweaver–Burke plots for one substrate at various constant concentrations of the other, result in a family of parallel lines (see Figure 27).

The physical meaning of the constants of the Alberty equation or of the Dalziel coefficient is obtained by solving the steady-state equations for different mechanisms and obtaining expressions for each K or ϕ in terms of

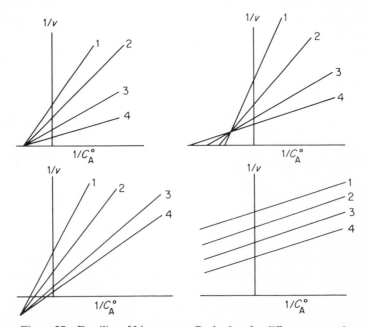

Figure 27   Families of Lineweaver–Burk plots for different cases of two substrate reactions discussed in the text. In each case the lines 1 to 4 correspond to measurement at increasing concentration C_B°, but C_B° is constant for the measurement represented by any one line

rate constants for the individual steps of the mechanism (see, for instance, Dalziel and Dickinson, 1966 and Dalziel, 1969). As more and more complex mechanisms are examined the solution of the matrix of simultaneous equations becomes more tedious. It also becomes progressively more of a purely algebraic exercise since only a limited number of steps can be identified by steady-state kinetics. The method of King and Altman (1956) uses a schematic procedure for solving the $n - 1$ steady-state equations which describe a mechanism with n forms of the enzyme. It is also necessary that the equation for the conservation of the concentration of enzyme sites holds.

The effects of added product on the initial velocity of disappearance of substrate are a valuable clue to the identification of the mechanism. Some of the comments made here might be understood better after study of Section 6.5.1, where competitive and non-competitive inhibitions are analysed.

If one considers an ordered mechanism with A adding first to the enzyme and Y coming off first one would expect A and X to compete for the same site. In such a mechanism X is a competitive inhibitor for A. On the other hand, if one considers further the reaction in that direction $(A + B \rightarrow X + Y)$ in the presence of increasing concentrations of Y the following situation arises. As EX is formed it will become more and more likely for EXY to be formed before X can dissociate. So Y will be an inhibitor but not a competitive one. In the case of a random-order rapid equilibrium mechanism X will be a competitive inhibitor for A and Y will be a competitive inhibitor for B and vice versa for the reverse reaction. Similar arguments can be used for substrate or 'dead-end' inhibition. If one studies the effect of the ethanol concentration on the initial velocity of the alcohol dehydrogenase reaction

$$\text{Ethanol} + \text{NAD}^+ \rightleftharpoons \text{Acetaldehyde} + \text{NADH} + \text{H}^+$$

one finds that there is an optimum concentration above which ethanol inhibits the reaction. Alcohol dehydrogenase operates through an ordered mechanism and the inhibition by excess ethanol is due to the formation of $E_{\text{Ethanol}}^{\text{NADH}}$ after the aldehyde has dissociated from the ternary complex. In keeping with the sequence of events for an ordered mechanism the rate of dissociation of NADH is inhibited by the combination of ENADH with ethanol.

The rate of dissociation of NADH from the complex $E_{\text{Ethanol}}^{\text{NADH}}$ or $E_{\text{Aldehyde}}^{\text{NADH}}$ is not necessarily zero, it is just very much slower than the rate of dissociation from E NADH. This applies to many statements about mechanisms. Unless numbers for rate constants of alternative pathways are quoted, the preference of one pathway over another should not be accepted as absolute. When effects of changing conditions on enzyme reactions are examined particular attention has to be paid to the possibility that gradual changes in the balance between different mechanisms might occur.

6.4.4. Isotopic Exchange Reactions

The use of radioactive tracers to map out the pathways of metabolic sequences as well as of single enzyme reactions, together with chromatographic techniques for the separation of intermediates, have provided most of the contributions to qualitative biochemistry during the last twenty years. In addition much quantitative information has been obtained from studies of the rates of incorporation of labelled groups. Apart from some special investigations the following radioactive isotopes are most extensively used:

Isotope	Halflife	Maximum energy (meV)
H^3	12·3 years	0·018
C^{14}	5,570 years	0·155
S^{35}	87 days	0·165
P^{32}	14·2 days	1·71

Methods for the exploration of the stereospecificity of enzyme reactions, of points of fission and of exchange with the solvent are treated in some detail elsewhere (Gutfreund, 1965). In this section a brief summary will be presented of the method of interpreting the rates of isotope exchange of reactions at equilibrium. No attention will be paid to the isotope rate effect until this phenomenon is discussed in relation to the determination of rate limiting elementary steps in 7.4.1.

The use of exchange rates at equilibrium for the distinction between ordered, random rapid equilibrium and substituted enzyme mechanisms was introduced by Silverstein and Boyer (1964). Some specific examples will illustrate the potentialities of the procedure. The various possible mechanisms and exchange reactions suggest such a large number of permutations and combinations of experiments that no attempt is made to give even a fair selection. Hopefully the suggestions made will help to design experiments for each special case.

The use of isotope exchange studies for the distinction between the two sequential mechanisms mentioned (ordered addition or random rapid equilibrium) is particularly valuable. If the reaction

$$\text{Alcohol} + \text{NAD}^+ \rightleftharpoons \text{Aldehyde} + \text{NADH} + \text{H}^+$$

is at equilibrium a number of exchange reactions can be observed. If C^{14}-labelled alcohol is added it can exchange with aldehyde, C^{14}-labelled NAD can exchange with NADH and tritium-labelled $NADH^3$ can exchange with alcohol. There can be no exchange between NAD and aldehyde. If one wishes to study the concentration dependence of the rate of any of these exchange processes at equilibrium one has to change the concentration of two reactants, one on each side of the reaction. In the random rapid equilibrium mechanism the rate-limiting step is the interconversion of the ternary complex and the exchange from NAD^+ into NADH proceeds at about the same rate as the exchange in between the other pair of substrates. In the case of horse liver alcohol dehydrogenase with ethanol as substrate the rate of dissociation of NADH is the rate-limiting process and the nucleotide will not dissociate, to an appreciable extent, from the ternary complex. If one added labelled NADH to the equilibrium mixture the rate of exchange with either NAD^+ or ethanol would be slower than if labelled aldehyde were added to exchange with ethanol. Increasing concentration of aldehyde and ethanol will slow down the exchange of NADH.

While for exchange in sequential mechanisms all four reaction partners must be present with the enzyme, in the case of substituted enzyme mechanisms exchange can be observed, at times, with only two out of four reactants. For instance, if a catalytic amount of the enzyme transaminase is added to a mixture of either aspartate and oxaloacetate or glutamate and α-oxoglutarate, no net reaction will take place. However, if some C^{14}-labelled aspartate or glutamate is added to the respective mixtures, there will be an exchange of the carbon skeleton between the reaction partners. In this, as in many other situations described by Gutfreund (1965) and Rose (1966), the exchange pattern gives information about the mechanism without the use of rate measurement.

6.5. REVERSIBLE EFFECTORS OF ENZYME REACTIONS

6.5.1. Competitive and Non-Competitive Inhibition

Kinetic studies of reversible inhibition phenomena are used to obtain information about the conformation and reactivity of the active site. A reversible inhibition involves an inhibitor which forms an enzyme compound in equilibrium with free enzyme and inhibitor

$$E + I \rightleftharpoons EI$$
$$K_i = C_E C_I / C_{EI}$$

Competitive inhibitors are those which compete with the substrate for the possibility of binding to the enzyme. Enzyme–inhibitor–substrate ternary complexes cannot be formed with competitive inhibitors.

The effects of a competitive inhibitor on the apparent parameters of an enzyme–substrate reaction can be derived from steady-state equations.

For the simple system

$$E + S \underset{k_{-1}}{\overset{k_1}{\rightleftharpoons}} ES \overset{k_2}{\rightarrow} E + P$$

$$E + I \overset{k_3}{\rightleftharpoons} EI$$

The conservation of enzyme sites gives

$$C_E + C_{ES} + C_{EI} = C_E^\circ$$

and the steady-state equations are

$$dC_{ES}/dt = k_1 C_S[C_E^\circ - C_{ES} - C_{EI}] - [k_{-1} + k_2]C_{ES} = 0$$

$$dC_{EI}/dt = k_3 C_I[C_E^\circ - C_{ES} - C_{EI}] - k_{-3}C_{EI} = 0$$

$$dC_P/dt = v = k_2 C_{ES}$$

On eliminating C_{EI} from the two steady-state equations and substituting v/k_2 for C_{ES}, V for $k_2 C_E^\circ$ and K_m for $(k_2 + k_{-1})/k_1$ (see Section 6.3.4) the following equation is obtained for the initial velocity at a given substrate and inhibitor concentration:

$$v = \frac{V C_S}{K_m(1 + C_I/K_i) + C_S} \tag{6.47}$$

Comparison of this equation with equation (6.23) shows that the apparent K_m in the presence of a competitive inhibitor at concentration C_I is

$$K_m^I = K_m(1 + C_I/K_i)$$

The test for a competitive inhibitor is that the maximum velocity attained at infinite substrate concentration is unchanged.

The determination of K_i for competitive inhibition provides thermodynamic information about interaction at the binding site since it involves an equilibrium process.

A non-competitive inhibitor is defined as one which can form ternary complexes with enzyme and substrate, but these are inactive complexes

$$K_i = C_E C_I/C_{EI} = C_{ES}C_I/C_{IES}$$

$$E \rightleftharpoons ES \rightarrow E + P$$
$$S + I + \updownarrow \qquad \updownarrow$$
$$EI \rightleftharpoons IES \nrightarrow$$

Using the steady-state procedure to derive the expression for the initial velocity of an enzyme reaction in the presence of a given substrate and non-competitive inhibitor concentration one obtains

$$v = \frac{VC_S/(1 + C_I/K_i)}{K_m + C_S} \tag{6.48}$$

This shows that a non-competitive inhibitor has no effect on the apparent K_m but the apparent maximum velocity V^1 at concentration C_I is

$$V^1 = V/(1 + C_I/K_i)$$

Figure 28 shows the effects of different concentrations of competitive and non-competitive inhibitors on Lineweaver–Burke plots. The various graphical

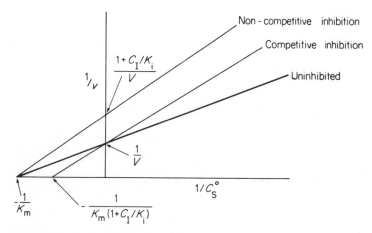

Figure 28   The effects of competitive and non-competitive inhibitors
on Lineweaver–Burk plots

methods for the evaluation of K_i for both types of inhibition are discussed by Dixon and Webb (1964).

Uncompetitive inhibition is defined as inhibition by a compound which can only combine with an enzyme–substrate complex and not with the free enzyme. In the presence of an uncompetitive inhibitor both the apparent K_m and the apparent V are altered.

It should be emphasized that a number of more complex forms of inhibition can occur. Mixed inhibition will be frequently encountered. This is most easily visualized when the topic of the next section has been considered. Protons can often be identified as largely non-competitive inhibitors, yet their presence at the active site will affect substrate affinity and the presence of the substrate will affect the affinity for the proton. It should also be noted that in a system with a K_m as described by equation (6.41) and $k_3 \gg k_2$, an inhibitor affecting k_3 will be uncompetitive.

Another complexity to be considered is the possibility of partial inhibition. This is an interesting phenomenon which, like minor alternative pathways, is often overlooked because of insufficient accuracy of the data. Tables of expressions for apparent K_m^I and V^I for different forms of complex inhibition are given by Dixon and Webb (1964) and Gutfreund (1965).

6.5.2. Hydrogen Ions as Modifiers of Enzyme Kinetics

The most common pursuit among enzymologists, apart from determinations of K_m and V values, is to investigate the pH dependence of the rate. From the latter deductions are often made about the nature of some particular amino acid residue of the enzyme which takes part in the catalysis. While such investigations have often provided correct answers, for instance about the role of histidine in the reactions of chymotrypsin and trypsin (Hammond and Gutfreund, 1955), a facile interpretation is full of pitfalls. Some guidelines can be provided to show how pH effects can be investigated. Protons are best considered as 'modifiers'. They can act both as inhibitors and as activators of enzyme reactions.

In the present discussion effects of pH on the ionization of the substrate will not be considered. In practice, one always has to be aware of possible effects on the ionizing groups of the substrate. This is usually not difficult to take into account. The next point which has to be considered and disposed of is that one must be sure that all effects of pH changes are reversible. Many enzymes are sensitive to extremes of pH and readily undergo irreversible changes when exposed to them.

In the case of trypsin catalysed hydrolysis reactions the pH dependence of V (see Figure 29) is represented by a simple ionization curve of a single group; the enzyme being in the active form when that group is in its basic form. More usually one finds bell-shaped curves representing the pH dependence. For this purpose it is essential to ascertain extrapolation to true substrate saturation over the whole pH range. For the present the following scheme will be considered to interpret the data:

$$S + EH_2 \rightleftharpoons EH_2S \rightleftharpoons EH_2P \rightleftharpoons EH_2 + P$$
$$\updownarrow \qquad \updownarrow \qquad \updownarrow \qquad \updownarrow \qquad\qquad K_2^E = \frac{C_{EH}C_{H^+}}{C_{EH_2}}$$
$$S + EH \rightleftharpoons EHS \rightleftharpoons EHP \rightleftharpoons EH + P$$
$$\updownarrow \qquad \updownarrow \qquad \updownarrow \qquad \updownarrow \qquad\qquad K_1^E = \frac{C_E C_{H^+}}{C_{EH^+}}$$
$$S + E \rightleftharpoons ES \rightleftharpoons EP \rightleftharpoons E + P$$

There will be two other pairs of dissociation constants to describe the ionization of the enzyme substrate and product compounds. At substrate saturation the pH dependence will be a function of the dissociation constant

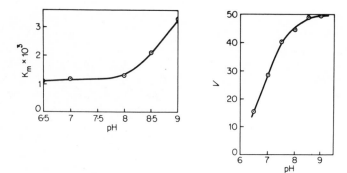

The pH-dependence of K_m and V for the chymotrypsin catalysed hydrolysis of acetyl L-phenylalanine ethyl ester at 25°. The marked increase in K_m at alkaline pH made it difficult to determine the Michaelis parameters above pH 9. This lead to erroneous data indicating a decrease in the maximum velocity at alkaline pH

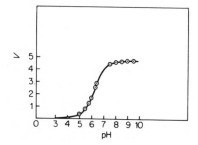

The pH-dependence of the maximum velocity of the trypsin catalysed hydrolysis of benzoyl L-arginine ethyl ester at 25°

Figure 29   The pH dependence of Michaelis parameters of chymotrypsin and trypsin catalysed reactions

of that intermediate which precedes the rate-limiting step. For the present discussion it is assumed that EHS → EHP is the rate-limiting step and that ES and EH$_2$S are either catalytically inactive or cannot be formed. Such a system will have a bell-shaped pH activity profile:

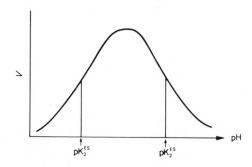

In Figure 30 it is illustrated that the pH profile can give apparent pK values which bear no relation to the ionizing groups at the catalytic site if the changes in pH result in a change in rate-determining step.

At times one finds statements in the literature to the effect that at a particular pH a certain fraction of the enzyme is inactive. This is not correct and wastes useful information. The equilibrium reactions between free

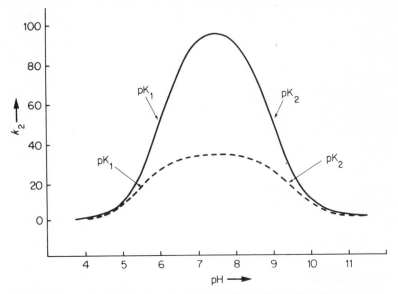

Figure 30   A simulated representation of the pH profile of an enzyme reaction in which two steps have rate constants (k_2 and k_3) of comparable magnitude. k_2 is pH-dependent as indicated by the solid line (k_2 optimal is 100 sec^{-1}). k_3 is pH-independent and has the constant value of 50 sec^{-1}. The pK values for the pH-dependence of k_2 and the apparent pK values for the overall rate are indicated on the diagram

protons and the different ionic forms of the enzyme are very much faster than any of the other steps of interconversion during catalysis. The protons are likely to go on and off a good many times say during the interconversion of the enzyme–substrate complex to enzyme–product complex. It is, therefore, valid to say that all the enzyme is active part of the time. The consequence of this statement is that the hydrogen ion equilibrium affects a rate constant. With this in mind one can argue that if the K_m does not change over a pH range which causes marked changes in V, the K_m is determined by rate constants prior to those of the rate-limiting step.

The maximum velocity at a given pH, V^{H^+}, is related to the true maximum velocity V under the conditions (sometimes unattainable) when all the enzyme and its compounds are in the right ionic form for activity, by the following equation:

$$\frac{V}{V^{H^+}} = 1 + \frac{C_{H^+}}{K_2^{ES}} + \frac{K_1^{ES}}{C_{H^+}} \tag{6.49}$$

The proportion of active enzyme EHS is given by

$$\frac{C_{H^+}}{K_1^{ES}} = \frac{C_{EHS}}{C_{ES}} \quad \text{and} \quad \frac{K_2^{ES}}{C_{H^+}} = \frac{C_{EHS}}{C_{EH_2S}}$$

From these equations one can evaluate graphically the two dissociation constants of the groups involved in catalysis by using the data from the two limbs of the bell-shaped pH/activity profile. Plots of $1/V^{H^+}$ against C_{H^+} have slope $1/K_2^{ES}$ and plots of $1/V^{H^+}$ against $1/C_{H^+}$ have slope K_1^{ES}; provided K_1^{ES} and K_2^{ES} are well separated. K_2^{ES} and K_1^{ES} can also be calculated from the hydrogen ion concentrations, $C_{H^+}^2$, at which half maximum velocity is reached on the acid side of the pH maximum and $C_{H^+}^1$ at which half maximum velocity is reached on the basic side of the pH optimum:

$$K_2^{ES} = C_{H^+}^2 + C_{H^+}^1 - 4\sqrt{(C_{H^+}^2 C_{H^+}^1)}$$

$$K_1^{ES} = C_{H^+}^2 C_{H^+}^1 / K_2^{ES}$$

The equation for the relation between K_m for the active form EH and K_1^E and K_2^E from the pH-dependence of $V^{H^+}/K_m^{H^+}$ dissociation constants for E and ES:

$$\frac{K_m^{H^+}}{K_m} = \frac{1 + \dfrac{C_{H^+}}{K_2^E} + \dfrac{K_1^E}{C_{H^+}}}{1 + \dfrac{C_{H^+}}{K_2^{ES}} + \dfrac{K_2^{ES}}{C_{H^+}}}$$

The denominator of this equation is equal to V/V^{H^+} and one can evaluate K_1^E and K_2^E from the pH-dependence of $V^{H^+}/K_m^{H^+}$

$$\frac{V^{H^+}}{K_m^{H^+}} = \frac{V/K_m}{1 + \dfrac{C_{H^+}}{K_2^E} + \dfrac{K_1^E}{C_{H^+}}}$$

If two ionizing groups of the enzyme are involved plots of pH against $V^{H^+}/K_m^{H^+}$ will give bell-shaped curves and K_1^E and K_2^E can be determined graphically by plotting the data of the acid and basic arm as for the determination of K_1^{ES} and K_2^{ES} indicated above.

Dixon and Webb (1964) describe other graphical procedures and give many examples of investigations into the pH-dependence of enzyme reactions.

One classical investigation illustrating an additional point is the study of the pH-dependence of the fumarase catalysed reaction (fumarate $\rightleftharpoons$ malate) in both directions. Frieden and Alberty (1955) have evaluated pK_1 and pK_2 for free enzyme, the enzyme fumarate complex EF and the enzyme malate complex EM. The pK values for free fumarate (F) and free malate (M) are included here in brackets.

	E	EF	(F)	EM	(M)
pK_1	6·8	7·3	(4·40)	8·4	5·11
pK_2	6·2	5·3	(3·02)	6·6	3·48

By the law of microscopic reversibility, the same groups on the enzyme must be involved in both directions. The different environment in the EF and the EM complex can result in the different pH profile of the reaction in the two directions. An important point which can be discussed in terms of the reaction

$$F + E \underset{}{\overset{k_1}{\rightleftharpoons}} EF \underset{}{\overset{k_2}{\rightleftharpoons}} EM \underset{}{\overset{k_3}{\rightleftharpoons}} E + M$$

is the uncertainty of the assignment of the pKs to EF or EM. It has been pointed out above that K_1^{ES} and K_2^{ES} refer to the particular enzyme complex which precedes the rate-limiting step—if there is a distinctly rate-limiting step. If the rate-limiting step for the hydration of fumarate is k_2 and the rate-limiting step for the reverse reaction k_{-2}, then the above assignment is correct for the different pKs of EF and EM. If, however, the rate-limiting steps in the two directions were k_3 and k_{-1}, respectively, then the assignment of pKs should be reversed. Whatever turns out to be the true situation in the case of fumarase; proper characterization of pKs of particular enzyme intermediates is a very important part of establishing catalytic pathways. This detail can only be obtained if individual steps are studied by transient kinetic techniques (8.2.1).

6.5.3. Metal Ion Activation and Inhibition

The direct and indirect participation of metal ions in enzyme catalysis has many facets. In this section we are concerned primarily with one set of examples, the role of divalent ions like Mg^{2+}, Ca^{2+} and Mn^{2+}. These ions are in rapid equilibrium with their enzyme and/or substrate complexes. Similarly in the less direct influence of K^+, Na^+ or NH_4^+ ions on some enzymes, the ionic equilibria are fast compared with the turnover of the enzymes. The so-called metallo-enzymes represent quite a different phenomenon. In such systems iron, copper or zinc is quite firmly bound to the protein. Although it is often possible to remove the metal reversibly, the metal association–dissociation does not occur on the time scale of the turnover of the enzyme.

The role of small monovalent cations in their often very specific control of enzyme activity has many important biological implications (see for instance Sections 5.5.2 and 5.5.3). There is no doubt that protein molecules can distinguish between a sodium and a potassium ion and can respond specifically to contact with one or the other of them (Suelter, 1970). Very few quantitative treatments of such phenomena have been reported. Binding constants for activators can be evaluated from their concentration at half maximum activation or more precisely from a plot which makes good statistical use of the data.

The most widely investigated metal–enzyme–substrate complexes are those of the divalent cations with enzyme systems involved in phosphate transfer. These systems fall into three classes:

(1) No EM^{2+} complex in absence of S

$$M^{2+} + S + E \rightleftharpoons ES + M^{2+}$$
$$\updownarrow \qquad\qquad \updownarrow$$
$$SM^{2+} + E \quad \rightleftharpoons \quad ESM^{2+} \quad \longrightarrow \text{ Catalysis}$$

(2) No ES complex in absence of M^{2+}

$$M^{2+} + S + E \rightleftharpoons M^{2+}S + E$$
$$\updownarrow \qquad\qquad \updownarrow$$
$$S + EM^{2+} \quad \rightleftharpoons \quad EM^{2+}S \quad \longrightarrow \text{ Catalysis}$$

(3) No SM^{2+} complex in absence of E

$$M^{2+} + S + E \rightleftharpoons EM^{2+} + S$$
$$\updownarrow \qquad\qquad \updownarrow$$
$$ES + M^{2+} \qquad\qquad EM^{2+}S \quad \longrightarrow \text{ Catalysis}$$
$$\text{possible dead end}$$

Steady-state kinetics, relaxation kinetics and magnetic resonance techniques (Cohn, 1970; Mildvan, 1970) have been used extensively to distinguish between these three sequences. Yeast hexokinase, which catalyses the reaction

$$\text{Glucose} + \text{ATP} + Mg^{2+} \rightleftharpoons \text{Glucose 6-phosphate} + \text{ADP} + Mg^{2+}$$

can be used as an example to illustrate steady-state data which distinguish between models (1) and (2) above. Only the role of the magnesium ion is

considered here. Other aspects of the order of addition of the substrates have
been investigated by the methods discussed in Sections 6.4.3 and 6.4.4.

Gutfreund and Hammond (1963) have shown that under conditions
$C_{ATP} > C_{Mg^{2+}}$ free ATP binds to the enzyme with an affinity almost identical
to that of Mg ATP. Free ATP acts as a competitive inhibitor of Mg ATP.
Figure 31 illustrates the results to be expected for enzymes of types (1) and (2).
In enzyme systems of type (1), under conditions of $C_{ATP} > C_{Mg^{2+}}$, at constant
$C_{Mg^{2+}}$ the velocity of the reaction will be inversely proportional to C_{ATP}.
Under the same conditions for enzyme systems of type (2) the velocity of the

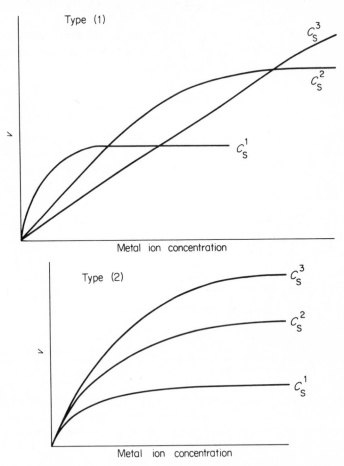

Figure 31   The metal ion concentration dependence of the
reaction velocity of reactions catalysed by type (1) and type (2)
kinases (see p. 151). The ATP concentration (C_S) is constant
for each curve and $C_S^1 < C_S^2 < C_S^3$

reaction will be either independent or proportional to C_{ATP}. The association between ATP and Mg^{2+} is discussed in Section 3.1.2.

Creatine phosphokinase and several other enzymes behave like yeast hexokinase. While a much larger group, pyruvate kinase among them, are of the metal bridge type (2). The magnesium requiring enzymes will also function with Mn^{2+} and Ca^{2+}, albeit much more slowly. The study of inhibition of Mg^{2+} activation by other ions clearly must make use of equilibrium binding data of the inhibitory ions with both substrates and enzymes. Mildvan (1970) has provided a detailed account of studies on metal requiring enzymes and a list of enzymes belonging to the different classes.

When metal ions are used to study inhibition phenomena with enzymes which do not require metals for their activities, the tools and interpretation are similar to those discussed in the preceding sections. The reactivity of specific groups apparently involved either in substrate binding or in catalysis can be explored by their interaction with inhibitory metal ions.

6.6. DEVIATIONS FROM MICHAELIS–MENTEN BEHAVIOUR

6.6.1. Substrate Activation or Inhibition of Monomeric Enzymes

Up to now it was assumed throughout this chapter that, however complex the detailed reaction mechanism discussed, the relation between initial velocity and substrate concentration in a single substrate reaction could always be expressed by

$$v = \frac{VC_S}{K_m + C_S}$$

For progressively more complex mechanisms the two constants V and K_m are composed of more and more individual rate constants. Adherance to the Michaelis mechanism is usually defined as giving linear reciprocal plots of the type given in Section 6.3.2.

In single substrate, monomeric (single site) enzyme systems many of the possibilities for non-linear substrate response of multi-substrate, oligomeric enzymes are excluded since there is, by definition, no interaction between binding sites. Rabin (1967) has suggested that the following mechanism could give rise to 'cooperative' substrate activation:

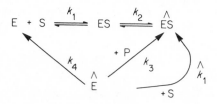

If k_2 is a slow isomerization of the enzyme–substrate complex, free enzyme being initially predominantly in the form E, then the first turnover after mixing E and S will be controlled by $k_2 C_{ES}$. After the first turnover there are alternate pathways. If $\hat{k}_1 C_S < k_4$ subsequent turnovers will proceed at the same rate as the first; if on raising the substrate concentration $\hat{k}_1 C_S > k_4$, the slow conformation change will be bypassed and substrate activation will occur. Similarly one can set up a complex reaction scheme in which the enzyme after the reaction is in a conformation which, if complexed with substrate, will result in a slower alternate pathway.

Monomeric single substrate enzymes can be activated or inhibited, respectively, by excess substrate if the free substrate complexes with an inhibitor or activator. If binding of the substrate to a second site, which is not a catalytic site, is included as a possibility the form ESS can be either more or less active.

Although there are few monomeric enzymes which catalyse two substrate reactions, there are some. Substrate inhibition of the following type is common:

EPB and EB are two types of several possible abortive complexes.

In all these cases substrate activation or inhibition can be used to obtain novel information about the mechanism.

6.6.2. Steady-state Kinetics of Cooperative Enzyme Systems

The kinetic analysis of cooperativity, interaction between binding sites, has already been referred to in Section 4.3.4. The steady-state velocity was used as a measure of the degree of site occupation by substrate. In the present chapter it has been shown that steady-state kinetics cannot always distinguish between substrate binding steps and catalytic phenomena. The elucidation of individual reaction steps from observations of single turnovers (see Section 8.2.2) will naturally help to describe any steps involved in cooperative interactions between sites. To save repetition only one additional example will be discussed here in any detail to illustrate the application of steady-state kinetics to the study of site interaction.

Before treatment of a particular system some general comments about the application of steady-state kinetics to the study of metabolic control should be useful. Cooperativity, involving sigmoidal response of activity to increasing substrate concentration, presents the possibility of a 'leaky switch' (see Figures 11 and 13). In a cooperative system a small change in substrate concentration at a critical level will produce a much larger change in activity than in a Michaelis type system.

A number of important consequences of cooperative phenomena have already been discussed in Chapter 4. These include the uptake and discharge of oxygen by haemoglobin over a small range of oxygen tension and a number of feedback inhibition phenomena controlled over small changes in metabolite concentration. Another aspect of effects of cooperative activators and inhibitors is of importance. In a single isolated enzyme reaction an inhibitor or activator cannot affect the equilibrium of the reaction. Many important biosynthetic pathways operate via different routes from the degradative pathway. Sometimes the cycles of biosynthesis are large and sometimes they are quite small. A simple cycle is illustrated and described in Figure 32. The two enzymes involved, phosphofructokinase and fructose 1,6-diphosphatase, are inversely controlled by ATP and other metabolites. ATP will inhibit the kinase above a certain concentration and it will activate the phosphatase. In this way in a cyclic process an effector can change the dynamic equilibrium of F6P and the diphosphate in a cyclic process. This

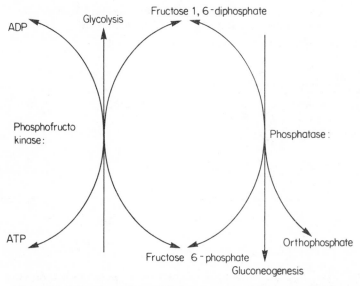

Figure 32 The dynamic equilibrium between fructose 6-phosphate and fructose 1,6-diphosphate can only be affected by activators and inhibitors when two enzymes are involved which respond in a different manner to the effectors. For instance, AMP and FDP activate the kinase and inhibit the phosphatase

is essential for the control of the balance of glycolysis and gluconeo-
genesis.

A good and quite different example of deviations from Michaelis kinetics
has been described by Engel and Dalziel (1969) in their paper on ox liver glutamate dehydrogenase. Figure 33 shows that plots of C_E^o/v against $1/C_{NADP^+}$
are not linear. Distinct linear regions were obtained from Lineweaver–
Burk plots. Going towards higher NADP$^+$ concentration each of a maximum
of four slopes indicated progressively higher activity and higher K_m. Engel
and Dalziel make it clear that this deviation can be due to several phenomena.
NAD$^+$ or NADP could bind to an effector site to activate the reaction, there
could be homotropic interaction between binding of nucleotide to successive
active sites and finally there could be interaction between active sites to
alter catalytic rate constants which contribute to K_m. The elucidation of the
mechanism will have to await detailed analysis by transient kinetic techniques.
However, one feature of the results on glutamate dehydrogenase is different
from other cooperative phenomena discussed so far. The interaction between
nucleotide binding sites, if interpreted in terms of binding, exhibits negative
cooperativity: successive binding sites become weaker. The model proposed
by Monod, Wyman and Changeux (see 4.3.6) cannot explain negative
cooperativity. The more general model derived by Koshland can result in
positive or negative site interaction. While it is as well to avoid calling the
Adair equation a model, it was designed to fit constants to a curve, the results
on glutamate dehydrogenase are compatible with it.

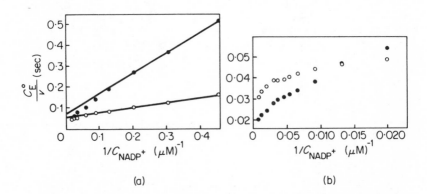

(a)   (b)

Figure 33   Variation of the initial rate of beef liver glutamate dehydrogenase
catalysed oxidation of glutamate with NADP$^+$ concentration. The reaction
was studied in phosphate buffer pH 8·0 at 25° in the presence of 50 mM
glutamate (○) and 3 mM glutamate (•). In (a) the range of C_{NADP^+} was from
2·0 to 50 μM and in (b) the range of C_{NADP_+} was from 50 to 1090 μM (from
Engel and Dalziel, 1969)

CHAPTER 7

Factors Affecting the Rates of Chemical Reactions

7.1. THE UPPER LIMITS OF REACTION VELOCITIES IN SOLUTIONS

7.1.1. Collisions, Encounters and Separation of Reactants

In a second-order reaction three events have to take place if the reaction is of the type

$$A + B \rightarrow C + D$$

and the first two of these have to take place in a reaction of the type

$$A + B \rightarrow C$$

(1) The reactants have to collide or to approach each other within reaction distance.
(2) The chemical transformation has to take place.
(3) The reaction products have to diffuse apart.

In the case of a first-order reaction the second of the above events has to take place for a reaction of the type

$$A \rightarrow B$$

and the second and third events have to take place for a reaction of the type

$$A \rightarrow B + C$$

For chemical reactions in solution it is important to consider the role of solvation. Figure 34 illustrates the distinct steps proposed by Eigen and de Maeyer (1963) for the reaction between 2-2 electrolytes. In this case, the third and rate-limiting step involves the removal of one water molecule between the two ions. In other reactions the two reactants break through each other's solvation shell and have an 'encounter'. Subsequent to this encounter, if the reaction probability is low, there can be several collisions before the reactants break out of their common solvation shell. The distinction between encounter and collision can be illustrated further by discussing the effects of viscosity changes on the two types of events. As the viscosity of the solvent is increased, encounters will be decreased because it

157

will become more difficult to break through the solvation shell. However, after the encounter the reactants will also find it more difficult to break out of their mutual shell. Reactants will have more collisions per encounter as the viscosity of the solvent increases. Increasing viscosity will decrease the

$$M^{2+} + SO_4^{2-} \longrightarrow MSO_4$$

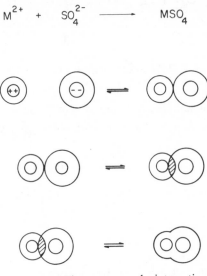

Figure 34   Three steps on the interaction of 2-2 electrolytes after Eigen and de Maeyer (1963)

encounter frequency but keep constant the collision frequency (encounters × collisions per encounter).

If a reaction occurs every time there is an encounter and subsequent collisions no longer contribute to the reaction, it is called a diffusion-controlled reaction. Consequently the rates of diffusion-controlled reactions are dependent on the viscosity of the solvent, while other reactions are not.

For biochemical processes the concept of diffusion control has other interest. Besides the relation between mechanism and diffusion control, numerical information about the maximum rate of a reaction is of importance for the control of some biological processes. Rates of reactions are much more important for their own sakes in the interpretation of systems (biological or chemical engineering) than they are in chemistry.

Another characteristic property of diffusion-controlled reactions is their relatively small dependence on temperature. The physical significance of the effect of temperature on chemical reactions is discussed in Section 7.2.2.

7.1.2. The Rates of Diffusion-controlled Reactions

In this section we are concerned with the problem of calculating the rate of a second-order reaction in a homogeneous solution for the special case that a reaction occurs every time there is an encounter of a pair of molecules of the two reactants. A related problem of importance for biological systems involves the calculation of diffusion from the site of production to the site of utilization in a heterogeneous or compartmented system. This latter aspect of diffusion control is closely dependent on the precise model for the system (Weisz, 1962).

For the case of uncharged molecules in solution the encounter rate constant k_e can be calculated from

$$k_e = \frac{4\pi N(D_A + D_B)(r_A + r_B)}{1,000} \ \text{M}^{-1} \text{sec}^{-1} \tag{7.1}$$

The units are those of a second-order reaction, N is Avogrado's number, and D and r are the diffusion constants and radii of the reactants A and B. For a system like an enzyme molecule with a radius of 30 A and a substrate molecule with a radius of about 5 A the encounter rate constant comes to approximately $5 \times 10^9 \text{ M}^{-1} \text{sec}^{-1}$. This can be compared with collision frequencies in a gas of $10^{11} \text{ M}^{-1} \text{sec}^{-1}$. In solution encounter frequencies somewhat greater than 10^{11} can be obtained for charged molecules.

The definition of a diffusion-controlled reaction can be somewhat amplified by saying that it occurs at every encounter with the appropriate relative orientation. The surface of a protein molecule with a radius of 30 A is 10,000 A^2, while the substrate molecule will only have 1/100 of that surface area. One would, therefore, expect the limiting rate of enzyme–substrate complex formation to be about $10^8 \text{ M}^{-1} \text{sec}^{-1}$. Even that would require that the substrate could encounter the specific fraction of the enzyme surface in any orientation. It is, therefore, remarkable that such a complex molecule as NADH combines with its site on a number of enzymes at rates of about $5 \times 10^7 \text{ M}^{-1} \text{sec}^{-1}$.

A number of examples will be discussed in Chapter 8 when methods for the determination of enzyme–substrate combination are described. It will be shown that in most cases the first step in enzyme–substrate combination is within the range of an order of magnitude (10^7–$10^8 \text{ M}^{-1} \text{sec}^{-1}$). Protein–protein interactions appear to be limited at rates of about $5 \times 10^5 \text{ M}^{-1} \text{sec}^{-1}$. These slower rates have been obtained for dimer to tetramer aggregation in haemoglobin, for actin–myosin combination, haemoglobin–antibody reactions and haemoglobin–haptoglobin combination, Kellett and Gutfreund (1970), Finlayson et al. (1969) and Noble et al. (1969).

The above semi-quantitative discussion should give a picture of the expected limits of second-order rate constants. This should be helpful in the planning and first interpretation of kinetic studies on association reactions.

7.1.3. The Upper Limit of First-order Reactions

The limiting rate of a first-order process of the type

$$A \xrightarrow{k_1} B$$

is best discussed by introduction of the idea of the transition state. If one considers that both reactant and product are in equilibrium with a structure X, called the transition state

$$A \rightleftharpoons X \rightleftharpoons B$$

then the equilibrium concentrations can be described by free energy differences in the following diagrammatic way:

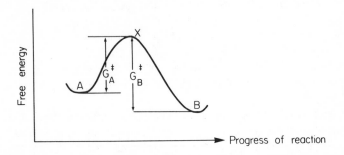

The equilibrium $\bar{C}_B/\bar{C}_A$ is equal to k_1/k_{-1} and it is also a function of $G_A^{\ddagger} - G_B^{\ddagger}$. From considerations of the relation between free energy and equilibria (1.3.4) it should be clear that

$$-\frac{G_A^{\ddagger} - G_B^{\ddagger}}{RT} = \ln k_1 - \ln k_{-1}$$

The rate of the reaction $A \to B$ is determined by

$$k_1 = \frac{kT}{h} e^{-G_A^{\ddagger}/RT} \tag{7.2}$$

where the term $e^{-G_A^{\ddagger}/RT}$ gives the fraction of A in state X and the term kT/h is the rate at which a molecule in the transition state is transformed into product. The model used in the transition state theory is that a molecule in that state is so weakened along the bond to be transformed that it will be changed at the next vibration. The frequency of vibration in one direction is given by

$kT/h =$ Boltzmann constant $\times$ Absolute temperature/Planck constant

$= 6 \cdot 25 \times 10^{12} \sec^{-1}$

These considerations show that the maximum rate of a first-order reaction, when $G^{\ddagger}$ approaches zero, is nearly $10^{13} \sec^{-1}$.

7.2. ABSOLUTE REACTION RATES

7.2.1. The Energy of Activation

In the previous section a thermodynamic quantity was used purely for a formal presentation of a model. In this section thermodynamic parameters are discussed in terms of reaction mechanisms. The aim of the theory of absolute reaction rates is to calculate the changes in thermodynamic parameters which take place during the reaction path from reactant to transition state and to product. From these thermodynamic parameters one can calculate the rate constants. Here we shall be concerned with the reverse programme, the determination of thermodynamic parameters from rate constants and deductions about the mechanism from these parameters.

Arrhenius found that over a moderate range of temperature there was a linear relation between the log of the rate constant of a reaction and $1/T$ and he expressed this as

$$\ln k = \ln A - \frac{B}{T} \qquad (7.3)$$

(T represents the absolute temperature which will be expressed as °K.) Equation (7.3) shows that there is also a temperature-independent term. The constant B is equal to E_A/R and

$$k = A \, e^{-E_A/RT} \qquad (7.4)$$

where E_A is the Arrhenius energy or activation energy. A, the pre-exponential factor, is related to the geometric restrictions of the reaction

$$A = PZ$$

Z is the collision frequency, encounters $\times$ collisions per encounter, and P is a steric factor. The contributions to the constant A have been discussed in Section 7.1.2. It will be shown in the following section how E_A is related to the free energy and heat of formation of the transition state or activated complex.

Arrhenius energies of enzyme reactions are frequently determined and included in a report on the study of an enzyme system, without any regard as to the value of such information. A single value for E_A under some arbitrary condition is quite useless for any deductions about reaction mechanisms. In some cases, of course, one may wish to know the temperature dependence of a reaction to interpret the response of a biological system to changes in temperature. The values of E_A for enzyme-catalysed reactions are usually near 10 kcal; exceptions are of interest and will be discussed below. It is of practical interest to consider the change in velocity when the temperature changes by 10°:

$$k_{25°}/k_{15°} = \exp(-E_A/R \times 298)/\exp(-E_A/R \times 288)$$

$$= \exp\left[\frac{E_A}{R}\left(\frac{1}{288} - \frac{1}{298}\right)\right]$$

if $E_A = 10,000$ cal and $R = 1.98$ cal deg.$^{-1}$, $k_{25°}/k_{15°} = 1.8$ and if $E_A = 11,000$ cal the ratio is 2. When the temperature is changed from 25° to 0° the rate constant is reduced by a factor of 4.75 (if $E_A = 11,000$ cal). This indicates that one cannot stop an enzyme reaction by putting the reaction mixture in the refrigerator.

Sometimes the temperature-dependence of an enzyme reaction has two components: the effect of temperature on the equilibrium between two forms of an enzyme (or substrate) and the effect on the equilibrium between ground state and transition state. An example of such a phenomenon is illustrated in Figure 35. The energy of activation of the trypsin-catalysed hydrolysis of benzoyl-L-arginine ethyl ester at pH 7.5 is 11.1 kcal and at pH 6 it is 18.2 kcal. Comparison with Figure 29 will show that at pH 6 the enzyme is in two forms and the activity will increase when imidazolium ion becomes less protonated at higher temperature (ΔH ionization 7 kcal). The picture as drawn for the condition pH 6.0 is oversimplified. There are only two points and more detailed information would show that, for such a case, the Arrhenius plot is not likely to be linear. Changes in slope occur when the reactants change with temperature (see 7.2.3).

7.2.2. Heat and Entropy of Activation

In Section 7.1.3 the free energy difference $G^{\ddagger}$ between the ground state and the activated state of a reactant was used to characterize the equilibrium between these states. The theory of absolute reaction rates is based on the assumption that this equilibrium has physical reality and that one can derive thermodynamic parameters for the transition state. A treatment analogous to the derivation of the van't Hoff equation (1.3.5) can be used to obtain $H^{\ddagger}$, the heat of activation, from the temperature-dependence of $G^{\ddagger}$ or of the equilibrium constant $K^{\ddagger}$ between ground state and transition state:

$$k = \kappa\frac{kT}{h}K^{\ddagger} = \kappa\frac{kT}{h}e^{-G^{\ddagger}/RT} \tag{7.5}$$

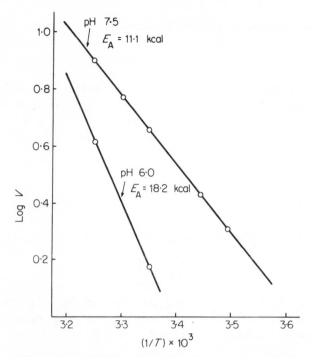

Figure 35 Arrhenius plots for the trypsin-catalysed hydrolysis of benzoyl L-arginine ethyl ester in 0·1 M NaCl and 0·01 M phosphate. (See discussion on p. 162)

κ is the transmission coefficient, the probability that a molecule in the transition state will give the product of the reaction. This is here assumed to be near unity and it will be omitted in subsequent equations.

Equation (7.5) can be expanded by use of the relations of equilibrium thermodynamics

$$k = \frac{kT}{h} \, e^{S^{\ddagger}/R} \, e^{-H^{\ddagger}/RT} \tag{7.6}$$

The slope of the Arrhenius plot of $\ln k$ versus $1/T$ is derived as follows:

$$\frac{d \ln k}{dT} = \frac{1}{T} + \frac{H^{\ddagger}}{RT^2}$$

$$-\frac{d(1/T)}{dT} = \frac{1}{T^2}$$

$$\frac{d \ln k}{d(1/T)} = \frac{d \ln k}{dT} \frac{dT}{d(1/T)} = \frac{-(H^{\ddagger} + RT)}{R} \tag{7.7}$$

The difference between the Arrhenius energy discussed in the last section and $H^{\ddagger}$ is

$$E_A = H^{\ddagger} + RT$$

At $25°$ $RT = 592$ cal, which is within the limits of error of most determinations of activation energies. The factor A in the Arrhenius equation is

$$\frac{kT}{h} e^{S^{\ddagger}/R}$$

From the relation

$$\left(\frac{\partial G^{\ddagger}}{\partial T}\right)_P = -S^{\ddagger} \qquad \text{(see Section 1.3.5)}$$

and equation (7.5) it follows that

$$\frac{\partial[T \ln(k/T)]}{\partial T} = \frac{S^{\ddagger}}{R} \tag{7.8}$$

and a plot of $T \ln(k/T)$ versus T will have the slope $S^{\ddagger}/R$. Volume changes $(V^{\ddagger})$ during activation to the transition change can be determined from the effect of pressure on the reaction velocity:

$$\left(\frac{\partial G^{\ddagger}}{\partial P}\right)_T = -V^{\ddagger}$$

$$\left(\frac{\partial \ln K^{\ddagger}}{\partial P}\right)_T = -V^{\ddagger}/RT$$

$$\ln k = k_0 - \frac{V^{\ddagger}}{RT} P \tag{7.9}$$

The interpretation of the changes in heat, entropy and volume during the activation process, in terms of reaction mechanisms, has attracted much attention but has achieved limited success. The overwhelming difficulty of interpreting thermodynamic parameters in terms of chemical change is due to the large effects of changes in solvation and of solvent structure. One of the important problems to solve for reactions of enzyme systems is the role of changes in the aqueous environment. In the study of relatively simple molecules interpretations in terms of changes in energy and in terms of distribution among a larger number of states (heat and entropy changes) can be successful if a large number of comparisons are made between a homologous series of compounds. Most predictions about reactivities and mechanisms in physical organic chemistry are based on a large amount of information of this kind. During the discussion of speculations about mechanisms of catalysis it will be seen that use is made there of contributions from entropy and energy changes. In the following section some additional complexities will be discussed, which arise in the interpretation of activation parameters of complex reactants such as enzymes.

7.2.3. Effects of Structure and Transitions

It was emphasized in Section 7.2.1 that Arrhenius plots are only linear over a limited temperature range and then only when a single observed reaction step is being changed by changing temperature.

One frequent result of changing conditions which change the rate of an enzyme-catalysed reaction is a change in rate-limiting step. An example rather similar to that discussed for the case of changing pH (Figure 35) is the following scheme in which reversal of steps is neglected:

$$E + S \underset{}{\overset{k_1}{\rightleftharpoons}} ES \overset{k_2}{\rightarrow} EP \overset{k_3}{\rightarrow} E + P$$

If the Arrhenius plots for the two rate constants k_2 and k_3 determined separately have the following form:

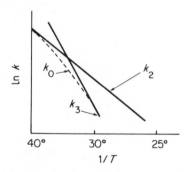

the overall rate measured as dC_P/dt will give the rate constant (see p. 135)

$$k_0 = \frac{k_2 k_3}{k_2 + k_3}$$

The plot of $\ln k_0$ versus $1/T$ will be curved going asymptotically into the line for k_2 above 40° and into the line for k_3 below 30°.

Another interesting phenomenon which results in changing slopes of Arrhenius plots is that due to changes in the structure of the reactants. In the example illustrated in Figure 35 there is a gradual change of E_A with pH due to the change in ionization of one group. A somewhat similar example is the rate of denaturation of a protein in the presence of urea. This is discussed in some detail by Tanford (1961). An actual inversion of the temperature-dependence occurs around 20°. The explanation of this behaviour is that there are two opposing effects. The rate of denaturation is proportional to the amount of urea bound to the protein but urea binding decreases as the temperature increases. Above 20° the increase in rate with increasing temperature can more than compensate for the decrease in bound urea.

A complex system of a different type is one which involves a sharp co-operative phase change. In Figure 36 the results of experiments with myosin are described. Myosin is one of the functional proteins involved in muscular

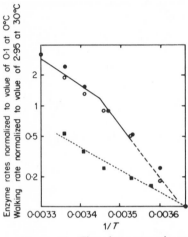

Effect of temperature on the ATPase activity of myosin and actomyosin and on the walking rate of ants. All enzymatic rates put on the same arbitrary scale, for comparison, by normalizing to a value of 0·10 at 0°C. Observed values, 0°C, for 'natural' isolated actomyosin (●) 0·0015 μmoles P_i/min/mg protein; reconstituted actomyosin (○) 0·0013 μmole P_i/min/mg protein; mycsin (■) 0·0010 μmole P_i/min/mg protein. Curve for walking rate of ants (Solid line) was superimposed by normalizing to a value of 2·95 at 30°C. Actual measured rate of walking at 30°C was 3·05 cm/sec; at 9°C (lowest temperature studied) rate was 0·44 cm/sec.

Figure 36   The above experiments on the effects of temperature on myosin systems are taken from a paper by Levy, Sharon and Koshland (1959) which should be consulted for detail as well as for references to biological investigations on temperature effects

contraction and it also acts as an enzyme in its isolated form, when it catalyses the hydrolysis of ATP to ADP and orthophosphate. The behaviour of muscle and of isolated myosin changes very sharply at 17°, as indicated by the change in slope of the Arrhenius plot. Such sharp changes in slope cannot be explained by normal pre equilibria or by changes in rate-limiting steps. The results of Koshland (Levy et al., 1959) are best described by a cooperative transition or melting of myosin at 17°. The similarity in the relation between temperature and rate obtained for the walking rate of ants and the enzymic activity of actomyosin and myosin (with dinitrophenol) tempted Koshland to the conclusion that actomyosin controlled a rate-limiting step of muscle action.

Phase transitions in macromolecules are frequently encountered. Proteins and nucleic acids have sometimes quite sharp (cooperative) temperature dependent structural transitions. Such transitions will in turn have distinct effects on the temperature dependence of the reactions in which these macromolecules are involved.

Although pressure has so far been used much less extensively as a variable to study equilibria or rates, pressure-induced changes in pre-equilibria and phase transitions have been observed. It is quite likely that more extensive use of pressure as a forcing function to change chemical processes will open up a new source of information. This may prove to be especially useful for the elucidation of mechanisms of reversible interactions in solutions (see Section 8.3.1).

7.3. CATALYSIS

7.3.1. Elementary Theories of Catalysis

The formal statements that catalysts do not change equilibrium constants, do not enter into the stoichiometry of the reaction and accelerate the rate can all be amplified with additional formalisms and with discussions of chemical mechanisms.

One first has to consider how a catalyst can reduce $G^{\ddagger}$ of a reaction. It can do this either by lowering the free energy of the transition state but maintaining chemically the same pathway. The transition state is put into an environment which lowers its free energy, or the ground state is put into an environment which raises its free energy. This proposal includes the existence of additional complexes on the same pathway. The second possibility is for the catalytic pathway to be quite different and to go through a different transition state of lower free energy. In a cyclic pathway

$$
\begin{array}{ccc}
A & \underset{k_1}{\rightleftarrows} & B \\
{\scriptstyle k_2}\big\updownarrow & & {\scriptstyle k_4}\big\updownarrow \\
AX & \underset{k_3}{\rightleftarrows} & BY
\end{array}
$$

all steps have to be in equilibrium and

$$\frac{C_B}{C_A} = \frac{C_{AX}}{C_A}\frac{C_{BY}}{C_{AX}}\frac{C_B}{C_{BY}}$$

The following reaction profile illustrates how the catalytic path through AX and BY progresses faster:

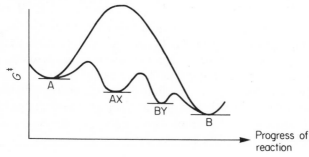

The distinction between the two modes of catalysis mentioned above is not always very clear and definitions of this kind are of limited value. The following quotation from Hinshelwood (1951) is very much to the point: 'There is no sense or profit in talking about theories of catalytic reaction in general. The theory of catalysis is the theory of chemical reaction velocity, and the methods of operation of catalysts are as diverse as the modes of chemical change'.

7.3.2. Homogeneous Catalysis

By definition, homogeneous catalysis is due to a catalyst in the same phase as the reactants. Heterogeneous catalysts are in a different phase either as a solid surface or a suspension of particles. Specially treated metal surfaces are the most common heterogeneous catalysts. Enzyme catalysis is probably best defined as homogeneous catalysis, although it has some of the physical characteristics of heterogeneous catalysis: adsorption and possible deformation of the substrate. The most pertinent examples of homogeneous catalysis for a discussion of the physical properties of enzymes are:

(1) Acid and base catalysis either due to the special action of H^+ or OH^- ions or general acid/base catalysis due to proton donors or acceptors.
(2) Nucleophilic catalysis due to a nucleophile which can form a transient intermediate with the reactants.
(3) Electrophilic catalysis, which can also involve the formation of transient intermediates.

Some physicochemical principles for the characterization of catalytic groups are discussed in Section 2.2.4 and detailed treatments of the chemical reactions which form the basis of homogeneous catalysis are given by Bruice and Benkovic (1966), Jencks (1969) and Bender (1970).

Catalysis by H^+ and OH^- ions is often not strictly in accordance with the rule that the concentration of the catalyst does not change during the progress of the reaction. For example, during the acid catalysed hydrolysis of acetamide

$$CH_3CONH_2 + H_2O + H^+ \rightarrow CH_3COOH + NH_4^+$$

hydrogen ions are used up. Normally, though not always, the change in catalyst concentration will be small compared with its total concentration. Other deviations from the strict definition of catalysts are due to poisoning, which is often caused by impurities. Enzymes are relatively labile and are broken down and resynthesized in biological systems. The different physical contributions which the three types of homogeneous catalyst mentioned above can make to the enormous acceleration of chemical reactions by enzymes are discussed in the next section.

7.3.3. Enzyme Catalysis

Enzymes are not only remarkably efficient catalysts but they are also very specific. The kinetic evidence discussed in Chapters 6 and 8 shows clearly that enzymes form a complex with their substrates and that the correct reactive complex is only formed with the right substrate. Four contributions to the acceleration of reaction rates by enzymes have been widely discussed:

(1) Proximity of the reactive part of the substrate, bound to the site of the enzyme, with two or three catalytic groups at the site is one of the multifarious consequences of the formation of the correct enzyme–substrate complex discussed below (see for instance the review of the reactions of ribonuclease by Barnard, 1969).

(2) Covalent catalysis is involved when, for instance, a nucleophile on the active site displaces a leaving group on the substrate. The enzyme–substrate bond is subsequently hydrolysed to form product and free enzyme. For efficient catalysis the nucleophile on the enzyme must be better than water and must be a better leaving group than the one it displaced (see survey by Bell and Koshland, 1971).

(3) General acid-base catalysis appears to be involved in most enzyme mechanisms—usually together with one of the other three features listed.

(4) Distortion of the substrate during binding can bring the conformation nearer to that of the transition state (see, for instance, chair to half-chair conformation change in lysosyme substrates, Blake et al. (1965).

The conformation change of the protein in response to contact with the right substrate has a different role from the change in substrate conformation. The so-called induced fit response (Koshland, 1960) of the protein results in a variety of contributions, which together could be responsible for the large rate enhancement. It is difficult to give a unique figure for the rate enhancement by an enzyme because of ambiguities of suitable comparisons. This is also the wrong question to ask. Before discussing the right question to ask, the formation of the enzyme substrate complex should be discussed in a little more detail.

Proximity alone, the localization of two reactants may only be equivalent to changing the standard state from molar to 55 molar. As part of or in addition to proximity comes a further factor. The necessary collisions to produce the right orientation can be reduced to each contact being successful, when the reactants are bound in the right orientation, will add to the rate enhancement by proximity. Two other factors, which will make a significant contribution, are the fact that three or even four catalytic groups and substrates can be brought together and that the whole solvent environment at the active site is uniquely determined by the structure formed by the protein and the substrate together. The environment at the active site can change the oxidation-reduction potential of a flavine group by the equivalent of 5 kcal (see Section 2.3.2)—a possible contribution of 10^4 to a rate.

With the availability of detailed pictures of the three-dimensional atomic structures of active sites and enzyme–substrate complexes through crystallographic studies, and of the temporal organization of the elementary steps through powerful new kinetic techniques, new questions should be asked. Can the knowledge gained from the rate of a single step in a known structure and a known environment teach us something new about reaction mechanisms? This, in the light of present-day information about enzymes, is probably more meaningful than the endless comparison of reactions of smaller 'simpler' molecules which are in fact less well-defined than the processes at the active sites of enzymes.

7.4. THE EFFECTS OF ISOTOPIC SUBSTITUTION

7.4.1. Theory of Isotope Effects on Rates and Equilibria

The use of isotopes to explore the pathways of reactions, by following the fate of a labelled atom, was briefly referred to in Section 6.4.4. Here we are concerned with the effects on reaction rates of substitution of chemically identical atoms of different mass. In investigations of reaction mechanisms the effect of substitution by a heavier atom can give information about the rate-limiting step of the reaction. If the substitution occurs at a bond which is cleaved during the rate-limiting step, there is a significant change in the reaction velocity. The size of the effect depends on the change of mass. For deuterium in place of hydrogen effects of up to 10-fold reduction in rates have been observed. For carbon the ratios are rather small:

$$\frac{k_{C^{12}}}{k_{C^{13}}} = 1 \cdot 25 \qquad \frac{k_{C^{12}}}{k_{C^{14}}} = 1 \cdot 5$$

For nitrogen

$$\frac{k_{N^{14}}}{k_{N^{15}}} = 1 \cdot 14$$

and for oxygen values of

$$\frac{k_{O^{16}}}{k_{O^{18}}} = 1 \cdot 19$$

are to be expected.

An elementary explanation of the physical basis of the kinetic isotope effect provides some further insight into the use of the theory of absolute reaction rates for the calculation of rate constants. More accurate and more realistic application of the theory requires the use of statistical mechanics, the relationship between properties of molecules and the macroscopic thermodynamic properties observed in the laboratory (Hammett, 1952;

Kauzmann, 1966; Hill, 1961). The description of isotope effects given below comes from the introduction to a review by Westheimer (1961) where a more sophisticated treatment can be found (see also Jencks, 1969).

As an example the discussion here is based on the discrimination between reactivity at a C–H and a C–D bond. There is no discrimination between the electronic, rotational or translational properties of H and D atoms. The vibrational frequencies observed by infrared spectroscopy are near the values of 2,900 cm^{-1} for C–H and 2,100 cm^{-1} for C–D. The zero-point energy of bond vibration ε is given by

$$\varepsilon = \tfrac{1}{2}h v \qquad \text{(h is the Planck constant and v the vibrational frequency)}$$

From Hooke's law the vibrational frequency is

$$v = \tfrac{1}{2}\pi\sqrt{(c/m)} \qquad \text{(c is the force constant of the bond and m is the mass of the atom)}$$

It follows that

$$\varepsilon = h4\pi\sqrt{(c/m)} \qquad (7.10)$$

The force constant of C–D and C–H bonds are essentially the same and the zero-point energy will be inversely proportional to the square root of the mass. Accordingly, the zero-point energy of C–D should be $1/\sqrt{2}$ times that of the C–H, which corresponds very closely to the relative energies calculated from the above infrared frequencies:

$$\varepsilon_{H-C} = 4\cdot15 \text{ kcal/mole} \quad \text{and} \quad \varepsilon_{D-C} = 3\cdot0 \text{ kcal/mole}$$

The following scheme indicates the relationship between zero-point energies and energy of activation. In the transition state the vibration is transformed into translation and there is no difference in the transition state energy of C–H and C–D:

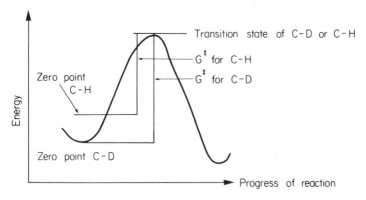

From the difference in $G^{\ddagger}$ for the two bonds (1·15 kcal/mole) one can work out the ratio of rate constants for reactions with their rate-limiting step involving breaking of the substituted bond. From equation (7.2) one can derive

$$\frac{k_H}{k_D} = \frac{e^{-G_H^{\ddagger}/RT}}{e^{-G_D^{\ddagger}/RT}} \qquad \cdot$$

$$\ln \frac{k_H}{k_D} = (G_D^{\ddagger} - G_H^{\ddagger})/RT = 1\cdot94 \text{ (at 298° absolute)}$$

$$k_H/k_D = 7$$

As will be seen in Section 7.4.2, isotope effects around this value are often found in reactions involving a deuterium-substituted carbon bond. The magnitude of the effect also depends on the force constant and in exchange reactions the heavier isotope will accumulate in the reaction partner with the larger force constant.

In many cases deuterium isotope effects are found to be considerably smaller than those calculated from the oversimplified theory given above. Westheimer (1961) discusses the reason for reduction of the primary isotope effect. When it becomes very small it becomes difficult to distinguish between primary and secondary effects. As a rough guide deuterium isotope effects smaller than 2 are likely to be due to secondary effects of substitutions on a step other than the bond-breaking one. These are discussed by Jencks (1969).

An important warning has to be included again about the difficulties of interpreting the meaning of isotope effects on enzyme reactions if there is some uncertainty whether a single-rate constant is responsible for the overall velocity measured. If two steps in the reaction have comparable rate constants, the apparent effect of isotopic substitution will be smaller than the true effect to be expected for that one of the steps which involves breaking of the substituted bond.

The study of the effects of substitution of C^{13} for C^{12} or O^{18} for O^{16} is, of course, much more difficult. The relatively small maximum effects listed above indicate that accurate rate measurements are needed. One interesting special procedure can be used if mass spectrometers with sensitivity for components at low concentrations are available. The natural abundance of different isotopes in the substrate and product can be measured. In this way one can determine whether the ratio of two isotopes of one atom of the substrate is different from that in the product (see Table 15), i.e. whether there is isotopic discrimination during the reaction.

Table 15. Natural abundance of stable isotopes of biological importance (from Nier, *Phys. Rev.*, **77**, 789 (1950))

Hydrogen	H^1	99·98
	H^2	0·02
Carbon	C^{12}	98·9
	C^{13}	1·1
Nitrogen	N^{14}	99·63
	N^{15}	0·37
Oxygen	O^{16}	99·8
	O^{18}	0·2

There are some difficulties in interpreting deuterium isotope effects if water is involved in the reaction and experiments in D_2O are carried out. The important role of the solvent in determining rates and mechanisms of reactions has been emphasized in several places. D_2O has very different properties from H_2O and marked solvent isotope effects are found. This is especially true with enzymes which are likely to undergo a number of minor but significant structure changes on transfer from H_2O to D_2O. The higher molar heat capacity, viscosity and temperature of maximum density of D_2O as compared with H_2O are all consequences of the even greater structural organization of heavy water.

Although the reaction path in the forward and reverse direction of an equilibrium process must go via the same transition state (free energy maximum), the isotope effect on the two rate constants need not be the same. This means that there can be an isotope effect on the equilibrium constant. As shown above, substitution by a heavier isotope at a bond involved in the rate-limiting step, will shift the equilibrium in favour of the compound with the larger force constant at the substituted bond. Whether there is an isotope effect on an equilibrium constant depends on the relative force constants of the bond broken and the bond made during the rate-limiting process. Inspection of equation (7.10) shows that the larger the force constant the larger the effect of isotopic substitution on the zero-point energy. As a result, in an equilibrium-exchange process the heavier isotope will accumulate in that partner of the equilibrium in which the bond with the larger force constant is formed.

Before discussing ionic equilibria in D_2O a practical point must be made about the measurement of pD. When a pH meter with a glass and a calomel electrode is calibrated with standard solutions of known C_{H^+} made up in H_2O and it is then used for the measurement of pD ($-\log C_{D^+}$) in D_2O solutions, the following relation holds

$$pD = \text{pH meter reading} + 0.4$$

For the dissociation equilibrium of an acid

$$CH_3COOD \underset{k_{-1}}{\overset{k_1}{\rightleftharpoons}} CH_3COO^- + D^+$$

one can assume that k_{-1} will not be markedly affected when D_2O is substituted for H_2O since this is a diffusion controlled process. The decrease in k_1 due to the deuterium isotope should be directly reflected in a decrease in the dissociation constant ($K = k_1/k_{-1}$) if D_2O is the solvent. This is in fact what is observed; acid dissociation constants in D_2O are usually about 1/5 those observed in H_2O.

7.4.2. The Use of Isotope Effects for the Elucidation of Mechanisms

Distinction between nucleophilic and acid/base catalysis by functional groups on enzymes is an important aspect of the elucidation of the catalytic mechanisms (see 2.2.4 and 7.3.3). Bender, Pollock and Neveu (1962) investigated the effect of H_2O and D_2O as solvent on general-base catalysis and on nucleophilic catalysis by simple imidazole derivatives. They found no isotope effect in nucleophilic catalysis. In general-base catalysis, which involves as an important step a proton (or D^+) transfer between water and base, k_{H_2O}/k_{D_2O} was found to be between 2 and 3. In several enzymes like chymotrypsin and cholinesterase, which are postulated to function through base catalysis by imidazole, isotope effects of 2–3 have been found in D_2O. The difficulties of interpreting solvent isotope effects, especially in enzyme reactions, have been referred to in the previous section. However, if one only used techniques which give unambiguous results for the study of enzymes, one's scope would be very limited indeed. Some ambiguities can be removed by suitable controls and corrections. One of these is the change in pK values on transfer from H_2O to D_2O.

Another problem to be considered is the stability of isotopic substitution towards exchange with the medium. For example in the reaction

$$NAD^+ + CH_3\overset{\overset{*}{H}}{\underset{H}{C}}-OH \rightleftharpoons NAD\overset{*}{H} + CH_3C\overset{H}{\underset{O}{\diagup}}\!\!\diagdown + H^+$$

catalysed by alcohol dehydrogenase, the transfer between the hydrogen atoms marked * of ethanol and NADH is completely stereospecific and they do not exchange with the solvent. The hydrogens of the alcoholic OH and H^+ are, of course, in isotopic equilibrium with the solvent. (For a detailed description of the stereospecificity of dehydrogenases see Gutfreund, 1965.) This means that kinetic studies of the deuterium isotope effect can be carried out on the one hand with specifically labelled deuteroethanol or NADD and another set of studies can involve the comparison of the reaction in H_2O

and D_2O. Shore and Gutfreund (1970) have determined the deuterium isotope effect of the hydride transfer process from ethanol to NAD^+ and have found it to be near 6. The procedures involved in the determination of individual chemical steps, such as the hydride transfer process, are discussed in Section 8.2.1. If one finds a significantly large isotope effect in a kinetic experiment, that is of course the evidence for having isolated a particular step in that experiment.

In hydrogen-transfer reactions catalysed by flavoproteins rapid exchange occurs between both the transferable hydrogens on the reduced flavine nucleotide and the solvent. The deuterium isotope effect has been studied for a number of reactions in which the transfer from specifically labelled substrate to enzyme-bound flavine can be followed distinctly from the rest of the reaction. For example, the transfer of deuterium from 1-deutero glucose to the flavine of glucose oxidase proceeds at approximately 1/8th of the rate of transfer of hydrogen (see Section 8.2.1).

Rose (1970) provided a valuable survey of the mechanisms of enzyme-catalysed proton abstraction. For instance, isotopically labelled dihydroxy-acetone phosphate can give information about conditions under which one is studying a particular step of the reaction of this substrate with either aldolase or triose phosphate isomerase. For more detailed advice on the interpretation of isotope effects in enzyme reactions the chapter on isotope effects in Jencks (1969) should be studied. Very few authors have recently concerned themselves with effects of isotopes other than those of hydrogen. Not only are the effects of deuterium or tritium replacement for hydrogen so much larger than for interchange of, for example, C^{13} for C^{12}, but one also has to consider the ease with which one can usually get isotopically pure deuterium derivatives. A classical paper which considers several aspects of the interpretation of small isotope effects in enzyme reactions (Seltzer, Hamilton and Westheimer, 1959) should be consulted if more insight into this problem is required.

CHAPTER 8

Transients and Relaxations

8.1. THE APPLICATION OF RAPID REACTION TECHNIQUES

8.1.1. Methods for Following Rapid Chemical Reactions

In conventional biochemical investigations the conditions of most experiments are so chosen that rate measurements are obtained on a time scale somewhere between 30 sec and 1 hr. Less than 30 sec requires special techniques for sufficiently accurate identification of zero time and more than an hour fatigues the operator and the unstable components of the reaction mixture. In this section and Section 8.1.2 we shall discuss methods for following processes within milliseconds or even microseconds from the initiation of the reaction. In Sections 8.1.3 and 8.1.4 we shall outline the reasons why it is essential to follow reactions with such time resolution.

Fifty years ago Hartridge and Roughton wanted to measure the rates of the reactions of oxygen and other ligands with haemoglobin. The conditions under which these rates could be observed were circumscribed both by the forward and reverse rate of the binding process. An irreversible second-order reaction can be slowed down indefinitely by using lower and lower concentrations of reactants. If the second-order rate constant of $A + B \rightarrow AB$ is $10^7 \text{ M}^{-1} \text{ sec}^{-1}$, and one measures the rate of the reaction started by mixing A and B at $C_A^\circ = C_B^\circ = 10^{-8}$ M, then the first halftime is 10 sec and can just about be measured by conventional means. The only problem would be whether the change in reactant concentrations can be measured sufficiently accurately, say to $\pm 10^{-10}$ M. Of all the physical techniques available for continuous monitoring of concentration changes, fluorescence is probably the most sensitive and this would barely detect changes of smaller than 10^{-9} M. In most reactions of real interest there arises the additional complication that the reaction is reversible and would not proceed to any significant extent if the reactants are below 10^{-6} M—what could be seen of the reaction would be dominated by the contribution of the dissociation process to the rate of reaching equilibrium (6.2.1).

Hartridge and Roughton (see Roughton, 1963) solved this problem by developing methods for following spectral changes and heat changes in reaction mixtures from approximately 1 msec after mixing the reactants. The principle of their procedure is illustrated in Figure 37. The two reactant solutions are driven under pressure through mixing jets into a capillary tube. The turbulent flow of the mixture down the tube at a speed of 10 m/sec permits observation at chosen time intervals by moving the monitor (photocell or thermocouple) down the tube. Details of the continuous flow technique and of the type of application for which it is essential are found in Roughton (1963) and Gutfreund (1969). While the method is still used for special purposes where the excellent resolution with respect to time and reactant concentration is essential, it is rather demanding on the supply of reactants.

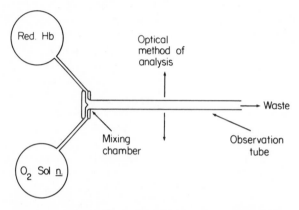

Figure 37   Schematic representation of the continuous flow method of Hartridge and Roughton for the measurement of rapid chemical reactions. The particular case illustrated here is the binding of oxygen to haemoglobin. The progress of the reaction can be observed at various points along the flow tube as the reaction ages. While optical observations are most commonly used, much useful information has been obtained by placing thermocouples or specific electrodes into the flow tube

For general application to the study of enzyme systems the so-called stopped-flow technique was developed by Chance and by Gibson (see Gibson, 1969). This method, which uses two small syringes to supply enzyme and substrate solution through a mixer into an observation chamber, is schematically illustrated in Figure 38. During flow the reaction mixture in the observation cell is between 1 and 3 msec old (depending on the optical path length in the chamber). When the stopping syringe hits the barrier, flow stops and a recording device is activated. Figure 39 shows the course of a dead-time calibration experiment. As can be seen, the stopping syringe actually triggers the oscilloscope a few milliseconds before flow stops.

The stopped-flow machine is now widely used for the study of rapid processes not only in enzyme systems but also in many other chemical

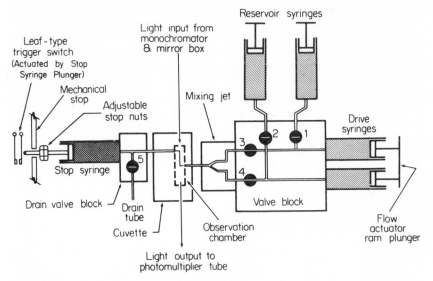

Figure 38   Schematic representation of the stopped-flow apparatus designed by Q. H. Gibson. This is reproduced by kind permission of Durrum Instrument Corporation from their Bulletin 131

reactions. The main developments of the equipment in recent years have been in the area of signal analysis such as the introduction of fluorescence detection and split-beam optics for the recording of absolute optical density changes. One of the major advantages of using the split-beam device illustrated in Figure 40 is that it gives accurate information about the amplitude of any very fast process which occurs during the dead time of the instrument. It will be seen that for many processes which can be observed, a rapid pre-equilibration can also be seen. This was found to be particularly applicable when pH changes monitored through the extinction change of indicators were used to investigate the nature of individual steps in enzyme reactions.

Any procedure which involves the adequate mixing of two liquids probably has a limit in time resolution of between 0·1 and 1·0 msec. When flow techniques are used at the limits of their time resolution, below about 2 msec, special care in the choice of the equipment is necessary and for studies on time scales of microseconds a different approach should be considered. If one has some means of perturbing a chemical system by subjecting it to a very fast pulse of energy one can follow a subsequent chemical reaction, which may be very fast, as long as it is significantly slower than the perturbing energy pulse. In a wider sense most applications of spectroscopy and fluorescence fall within the field of perturbation techniques, but the principles involved for such studies are beyond our present scope.

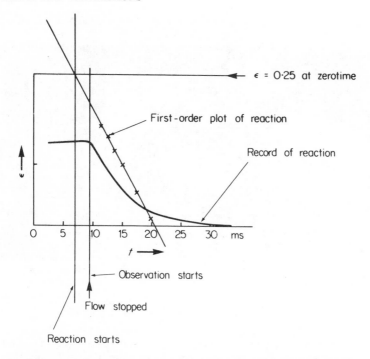

$\epsilon = 0.25$ at zerotime

First-order plot of reaction

Record of reaction

Observation starts

Flow stopped

Reaction starts

Figure 39   The dead time of a stopped-flow apparatus is the time taken for the reaction mixture to flow from the point of mixing through the observation chamber (dead time = time observation starts − time reaction starts)

The reaction record is due to mixing at 20° equal volumes of indophenol ($\varepsilon_{1\,\text{cm}}^{600} = 0.5$) in $0.5\,\text{M}$ phosphate pH 6 with $0.5\,\text{M}$ ascorbate in $0.5\,\mu\text{M}$ phosphate pH 6. This is a pseudo first-order reaction with a rate constant proportional to ascorbate concentration. The point in time at which the reaction starts is determined by extrapolating the first-order plot back to the time at which the extinction is the known extinction of the reaction mixture at zero time

If a system is subjected to a flash of light the energy levels of some molecules in the system are raised and a reaction can be initiated. A special case of such photochemical studies is the investigation of the rate and mechanism of the return to the initial stable state after flash photolysis. Properties of the triplet state of many molecules have been explored in this way (Porter, 1963). Examples of biological interest which demonstrate the power of laser flash techniques are illustrated in the work of Gibson and his colleagues on haem proteins. For instance, in the experiments into the recombination of haemoglobin and CO after photodissociation of the COHb complex,

Monochromator   Block containing mixer   Photomultipliers
                       and observation cells

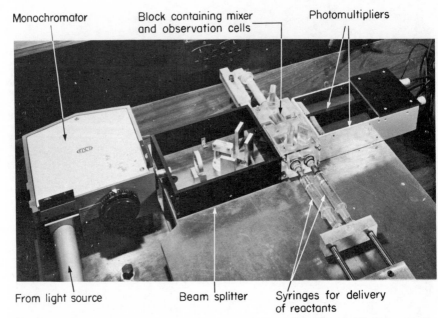

From light source   Beam splitter   Syringes for delivery
                                              of reactants

Figure 40   The split-beam stopped-flow apparatus developed by D. W. Yates, C. H. McMurray and H. Gutfreund to provide a record of rapid reactions in terms of absolute optical density changes. One light beam goes through two 0·5 cm cells, one from each solution, before mixing. The other light beam goes through a 1-cm observation cell containing the reaction mixture. The two beams fall on two separate photomultipliers. The output from the photomultipliers goes into an analog circuit which provides an output of $\log (I_0/I)$ for a storage oscilloscope

a combined stopped-flow-photoflash apparatus was used (Gibson and Parkhurst, 1968). In this way it is possible to cause a partial reversal of the reaction proceeding in the observation chamber by submitting it to a light flash at a chosen time after mixing. Some of the other perturbation techniques discussed in the next section can also be used to follow relaxation phenomena of steady states in a combined flow and relaxation device. However, only equilibrium relaxations will be discussed here.

8.1.2. Relaxation Techniques

Chemical equilibria are usually affected by one or more of the parameters temperature, pressure and electric field. The amplitude of the changes in concentration which occur when the temperature of an equilibrium system is changed by 10° depends on ΔH of the reaction (Section 1.3.5):

$$\left(\frac{\partial \ln K}{\partial T} \right)_P = \frac{\Delta H^\circ}{RT^2} \tag{8.1}$$

For a pressure change the amplitude of the change in equilibrium is given by

$$\left(\frac{\partial \ln K}{\partial P}\right)_T = -\frac{\Delta V^\circ}{RT} \tag{8.2}$$

The effect of an electric field on a chemical equilibrium with ΔM the difference in macroscopic dielectric moment between a mole of reactant and a mole of product is given by

$$\left(\frac{\partial \ln K}{\partial E}\right)_{T,P} = \frac{\Delta M}{RT} \tag{8.3}$$

There are a number of ways in which a solution can be heated by a few degrees in a few microseconds. Joule heating, microwave heating and laser flashes are some of the methods used for this purpose. With the most extensively used method, joule heating, a 5° rise in temperature can be obtained within 5 μsec if the ionic strength of the solution is reasonably high. If a reaction mixture at equilibrium at T_1 is suddenly heated to T_2 the concentrations of the reactants change to their new equilibrium position at the higher temperature. Figure 41 illustrates schematically the rapid heating process as indicated by the proton transfer from tris-buffer to indicator. The phenol red acts as an indicator of the change in C_{H^+} on heating. This type of experiment with a system of known heat of ionization can be used to calibrate the system in terms of temperature rise versus high voltage dis-

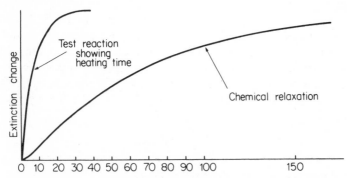

Figure 41   If a reaction mixture containing 2.5×10^{-5} M phenol red in 0·1 M tris-HCl O$_4$ buffer pH 8·2 is heated there will be a decrease in extinction at 570 mμ. Between 5° to 40° this extinction change is 0·0134 per °C for a 1 cm optical path. The reaction of the indicator is much faster than the heating time of a temperature jump apparatus with joule heating and the above record of the test mixture can be used to determine the heating time as well as the temperature change resulting from the electric discharge under particular experimental conditions

charge. The rate of equilibrium of the proton transfer in this reaction, at such high concentrations of reactants, is considerably faster than the record in Figure 41. What this record shows is the rate of heating and thus gives an indication of the time resolution of joule heating. Changes in some conditions can bring this down to about 1 μsec. The second trace on Figure 41 shows how a slower reaction would relax to the new equilibrium position.

The above comments should make it clear that a procedure which perturbs the ionization equilibrium also perturbs pH. The temperature-jump method, or the pressure-jump method discussed below, can be used to trigger a pH jump. Since many equilibria are pH-dependent, this offers another way of rapidly perturbing equilibria (see also temperature dependence of pH of buffer solutions, Section 2.2.1).

The amplitudes of displacements from equilibria at different reactant concentrations and at different temperature changes can be used to calculate equilibrium constants and thermodynamic parameters, respectively. The use of the pressure-jump technique also provides equilibrium parameters, in addition to kinetic information, and this should prove valuable in the elucidation of mechanisms. Only through a combined study of the effects of different forcing parameters, and the observation of the reactions by a range of monitors, can one obtain the full benefit of the powers of kinetic analysis.

The design of a cell for pressure-jump experiments is relatively easy for certain conditions. The optical observation of the reaction mixture provides sensitivity and versatility. Pressures below 200 atm (3,000 psi) are easy to contain and to release through a ruptured membrane within 50 μsec. The magnitude of expected changes in equilibria can be deduced from known volume changes in simple reactions. For reactions involving the formation of charged species in aqueous solution electrostriction of the solvent results in a change in volume of around -10 ml/mole. The percentage change in equilibrium constant $100(\Delta K/K)$ can be readily obtained by numerical substitution into

$$\left(\frac{\partial \ln K}{\partial P}\right)_T = -\frac{\Delta V^\circ}{RT}$$

For small pressure changes we can write

$$\Delta \ln K = -\Delta P \frac{\Delta V^\circ}{RT}$$

$$\frac{\Delta K}{K} = \frac{100 \text{ atm} \times 10 \text{ ml}}{82 \times 298} = 4 \cdot 1 \times 10^{-2}$$

($R = 82 \cdot 07$ ml atom mole^{-1} deg^{-1}).

Under the conditions of this example there would be a 4·1 per cent. change in equilibrium constant when the pressure is changed by 100 atm. From a practical point of view one wishes to know how large a signal one would get from such a perturbation of an equilibrium constant. This in turn makes one enquire into the amplitudes of concentration changes to be expected. In a process of the type $A + B \rightleftharpoons AB$ the relative concentration changes $\Delta C_A/C_A$, $\Delta C_B/C_B$ and $\Delta C_{AB}/C_{AB}$ will vary with the total concentration. The maximum change will be observed under conditions $C_A = C_B = C_{AB}$.

There are a number of ways in which pressure perturbations can be effected. The points which make the simple cell with bursting membrane most attractive for investigations of protein systems are as follows. While the pressure is slowly raised with a hand pump, readings can be taken of pressure versus some concentration variable of the system. This provides thermodynamic information prior to the occurrence of the relaxation process. It is also relatively simple to build other monitors (thermocouples, for instance) into the pressure cell. Optical observation can be very sensitive because the size of the cell permits the transmission of plenty of light. It is important to remember that the pressure release back to 1 atm results in the initiation of the kinetic measurements under conditions equivalent to those of experiments without pressure perturbation under normal laboratory conditions.

The relaxation techniques discussed so far have all relied on a single step-function perturbation. With temperature perturbation the reversal of heating is necessarily relatively slow and rapid successive pulsing would result in continuous heating. In the case of pressure or electric field perturbation the situation is different. Controlled frequency pressure fluctuation (sound absorption) or electric field fluctuation can be used to get information about the behaviour of dynamic systems with time constants in the fractional microsecond region. These are very interesting but rather specialised techniques (see Eigen and de Maeyer, 1963, and Kustin, 1969).

8.1.3. Elementary Steps in Enzyme Reactions

The range of turnover numbers of enzymes is very wide. Quite a number of enzyme-catalysed processes occur only at about 10 sec^{-1} per mole of active sites while some exceptionally fast ones, like carbonic anhydrase in the direction $H_2O + CO_2 \rightarrow H_2CO_3$, proceed with limiting rates of 10^6 sec^{-1}. Fortunately there are a large number of enzymes with turnover numbers (k_0) in the range $1 < k_0 < 500$ sec^{-1}. This, of course, means that the slowest step must have a rate constant corresponding—at least—to the turnover number and that individual steps must be measured with a resolution of milliseconds. Although some individual first-order rate constants in the microsecond range have been measured and their role in enzyme mechanisms elucidated, methods available at present are not sufficiently developed for the complete description of a mechanism with a limiting rate above 200 or $300 \sec^{-1}$.

Let us try and answer the question: what do we define as the mechanism of an enzyme reaction? C. A. Vernon suggested that the mechanism should be defined as the number and the structure of intermediates and the rate constants of the interconversion of these intermediates. The interconversions of identifiable intermediates (intermediate states which become significantly populated) are called elementary steps. Vernon's definition of mechanism specifically excludes the explanation of the catalytic process and of the profile of forces in it. This presents a sufficiently circumscribed question for us to have some hope of answering it with the methods available to us at this time.

X-ray crystallography, with some essential aids from chemistry, provides unique three-dimensional structures of enzymes in one or a few selected conformations. Kinetic observations complement the structural information by distinguishing a number of short-lived additional forms of the enzyme under investigation. It will also become evident in the further discussion of the application of rapid reaction techniques, that structural as well as purely kinetic information can be obtained. This is due to the fact that the observation of the transient formation and decomposition of intermediates provides the opportunity to determine their spectra or related properties. While steady-state kinetics has, in the right hands, provided essential basic information about the behaviour of enzyme systems, it is no substitute for the direct observation of transient intermediates.

If the simplest possible enzyme mechanism is taken to be one with a single enzyme substrate intermediate and a single enzyme product intermediate:

$$E + S \underset{}{\overset{k_1}{\rightleftharpoons}} ES \underset{}{\overset{k_2}{\rightleftharpoons}} EP \underset{}{\overset{k_3}{\rightleftharpoons}} E + P$$

then steady-state kinetics will only give information about that intermediate which preceeds the rate-limiting step. Often one step is quite clearly rate limiting and the enzyme accumulates all in one form in the steady state. Let us take the case $k_1 C_S \gg k_2$ and $k_2 = 10 k_3$, and consider the concentrations of the three forms of the enzyme as a function of time from mixing enzyme and substrate till the steady-state velocity of appearance of product is reached. In Figure 42 an analog computer simulation gives a diagrammatic presentation of the events during the pre-steady state or transient period and shows that, during this phase, direct kinetic information can be obtained about the rates of formation and decomposition of the intermediates which are not present at significant concentrations in the steady state.

The algebraic analysis and the design of kinetic experiments on transients of enzyme reactions are suitably divided into studies of transients in reactants (substrate and product) and transients in enzyme intermediates. In addition

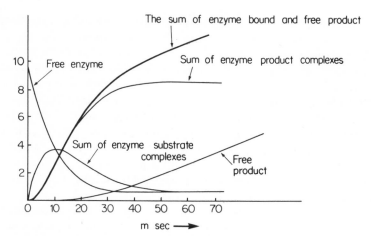

Figure 42   Computer representation of the reaction:
$$E + S \underset{}{\overset{k_1}{\rightleftharpoons}} (ES)_1 \underset{}{\overset{k_2}{\rightleftharpoons}} (ES)_2 \underset{}{\overset{k_3}{\rightleftharpoons}} (EP)_1 \underset{}{\overset{k_4}{\rightleftharpoons}} (EP)_2 \underset{}{\overset{k_5}{\rightleftharpoons}} E + P$$

The rate constants and concentrations assumed for this simulation were as follows: $C_S^o = 10^{-5}$ M, $C_{E1}^o = 10^{-7}$ M, $k_1 = 10^7$ M^{-1} sec^{-1}, $k_2 = 10^3$ sec^{-1}, $k_3 = 100$ sec^{-1}, $k_4 = 100$ sec^{-1}, $k_5 = 10$ sec^{-1} and
$$k_{-1}, k_{-2}, k_{-3}, k_{-4} < 10 \text{ sec}^{-1}$$

to the representations, drawn above, of the temporal changes in the three intermediates, one can draw the progress of product formation up to the straight line obtained in the steady state. During the pre-steady state period the record of product formation, enzyme bound and free, consists of the sum of exponentials. The analysis of such progress curves in terms of rate constants of elementary steps is given in some detail in Section 8.2.2. The analysis of transients of intermediates is an extension of the problem of solving the rate constants of consecutive processes (see 6.2.2). In Section 8.2.1 it will be shown how the observation of intermediates with distinct spectra can be used to complete the jig-saw puzzle of setting up the framework for some enzyme mechanisms.

The study of some of the elementary steps in enzyme reactions requires the use of rapid reaction techniques, but not of any kinetic equations in addition to those developed in Sections 6.1 to 6.2. A number of interesting phenomena can be used to investigate the events during enzyme–substrate combination separately from the catalytic or chemical interconversion steps. The study of partial reactions (8.2.3) and of some relaxation processes (8.3.1) is concerned with the elucidation of substrate recognition.

One of the interesting lessons derived from the information about elementary steps obtained so far is that one of three phenomena is found to be rate limiting for different enzyme reactions:

(1) A conformation change induced by initial contact between enzyme and substrate(s). This results in the formation of the complex with unique chemical reactivity.
(2) The chemical process(es) involved in the interconversion of the reactants: substrate(s) to product(s).
(3) The release of product, which is at times controlled by a necessary reversal of the substrate-induced conformation change.

Examples of different systems involving the above three elementary processes will be discussed in some detail in the next few sections. Finally, a summary of elementary steps should contain some comments about conformation changes involved in transitions between forms of an enzyme with different activities or substrate affinities. These steps are not on the obligatory pathway for the overall reaction and they are found to be perturbations of the normal pathway. Most transformations responsible for control phenomena are found to involve changes in affinity and can be studied in the absence of chemical interconversion. As will be seen (8.3.3), the conformational transformations between proteins with differing affinities occur on a wide range of time scales from several seconds to 100 μsec.

8.1.4. Rapid Biological Processes

There is a subtle but distinct difference in the use of kinetics as a tool for exploring reaction mechanisms and for describing biological processes. The point has been made in several places that the crucial question often is: how fast is a particular process or step in a process, under some defined condition? In the description of a biological system kinetics can be an end in itself, not just the means to an end, as it is usually regarded by chemists. Of course, in biology the control and optimization of systems are important considerations. Questions about which step controls a complex sequence of events and what would happen to the system if some specified step is accelerated or slowed down can only be answered with kinetic data. Most information about the rates of enzyme reactions has been obtained from measurements requiring the use of enzymes in solution at concentrations of about 0.1 μg/ml. Processes and phenomena were selected out to be studied over a period of a few minutes. Many enzymes occur in their natural environment at concentrations 10^4 to 10^5 times higher than those used for the enzyme assays mentioned above. If one wishes to determine the rate of steady-state flux through a sequence of steps, or if one wishes to observe the transient from one steady state to another, methods have to be used which permit the recording of the progress of the reaction on the time scale of milliseconds.

Enzymes in vivo or in well-reconstituted systems can differ in their dispersion from that found in an assay medium or a system designed for molecular weight determinations. First, many enzymes are membrane bound and behave quite differently when associated with lipoprotein than when they are in free homogeneous solution. Secondly, enzymes aggregate into polymers of their active units. This aggregation is controlled by protein concentration and other factors of the environment (see, for instance, 4.3.5). Thirdly, enzymes could interact with each other at high concentrations. Multi-enzyme complexes of firmly united molecules of different enzymes are well known (Reed, 1970). More speculative are suggestions that in homogeneous solutions of enzymes at high concentrations some interactions between successive enzymes might take place. The kinetic problems discussed in the rest of this section are intended to illustrate the use of rapid observations for the exploration of enzyme systems as distinct from the emphasis on reaction mechanisms developed later.

Kinetic studies on organized systems have stimulated the development of the most sophisticated techniques. In Chance's laboratory, equipment was set up in which pH, oxygen concentration, fluorescence and spectroscopic observations could be recorded simultaneously to monitor rapid processes occurring in mitochondria, cell suspensions or exposed live tissue. Although considerations of organized systems are beyond the scope of the present

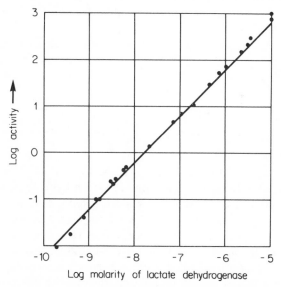

Figure 43   The catalytic activity in a solution of rabbit muscle lactate dehydrogenase is a linear function of the molarity of the enzyme (molecular weight 140,000 taken for calculation of molarity). The activity was measured as the rate of oxidation of NADH on addition of pyruvate at pH 6·4 (Wurster and Hess, 1970)

discussion, it is important to consider here how far they can be simulated when one starts with a number of preparations of pure enzymes. Rapid reaction techniques make it relatively easy to examine the relation between specific activity and enzyme concentration over many orders of magnitude. So far results obtained usually show the straight-line relationship of the experiments on rabbit muscle lactate dehydrogenase illustrated in Figure 43. At very low concentrations most enzymes lose activity. Any loss of activity due to adsorption on the wall of the reaction vessel and the presence of toxic trace metals will, of course, be more noticeable when there is a small total amount of enzyme. Apart from these effects there is the possibility of in-activation due to subunit dissociation. In enzyme systems where interesting protein concentration versus activity profiles have been found, such as glutamate dehydrogenase and phosphofructokinase, these are often due to dissociation into units with different affinities for effectors (see 4.3.5).

For the investigation of the contribution of structure to the function of a process of sequential steps, the study of transients can rapidly provide decisive information. In a biological system transients usually arise when the sudden release of substrate, or some activator, changes the steady-state rate. If we consider a sequential process $A \xrightarrow{\text{Step 1}} B \xrightarrow{\text{Step 2}} C$ and if we are re-cording the rate of appearance of C we would expect the following responses:

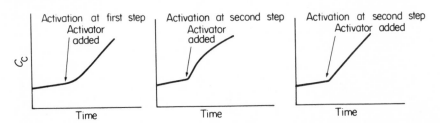

The response to activation at Step 2 will be sudden, as in the right-hand illustration, if it is either due to injection of B or if it is due to an activator and C_A is a large pool. If activation at the second step is due to addition of an activator and C_A is maintained at some steady-state concentration, then sudden activation will adjust itself as C_A falls to a new steady state (see middle illustration of responses). It has been suggested (Gutfreund, 1965) that the type of response obtained could be used to distinguish between some specific mechanisms. It should be pointed out that the distinction between sudden and gradual transitions is intended to be between transitions as rapid as a single turnover of the enzyme or the gradual change of the steady-state concentration of the intermediates.

The experiments of Hess and Wurster (1970) illustrated in Figure 44 clearly demonstrate the essential features of such an investigation. At zero

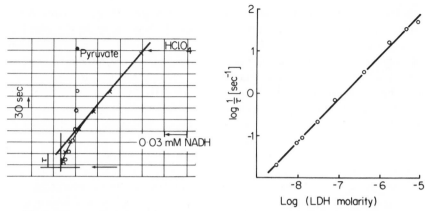

Figure 44   For the experiment illustrated on the left equal volumes of solution A (containing pyruvate kinase, lactate dehydrogenase, phosphoenolpyruvate, NADH) and solution B (containing pyruvate kinase, lactate dehydrogenase, adenosine diphosphate, NADH) in 50 mM imidazole-HCl buffer pH 6·4, 50 mM KCl and 8 mM Mg acetate at 25° were mixed to initiate the sequential reactions:

$$\text{Phosphoenol pyruvate} + \text{ADP} \rightleftharpoons \text{Pyruvate} + \text{ATP}$$

$$\text{Pyruvate} + \text{NADH} + \text{H}^+ \rightleftharpoons \text{Lactate} + \text{NAD}^+$$

The fine line shows the recorder trace of decrease in extinction at 340 mµ (representing oxidation of NADH in the second reaction). The heavy line is the extrapolation of the steady-state rate and $\bigcirc$ and $\times$ are the calculated values for pyruvate and NADH concentrations, respectively. The point $\bullet$ is the assayed value for the steady-state concentration of pyruvate after the reaction was stopped by the addition of HClO_4.

On the right the plot of $\log 1/\tau$ against $\log[\text{LDH molarity}]$ shows that the transient time is proportional to LDH concentration. Both enzymes were prepared from rabbit muscle (Hess and Wurster, 1970)

time the concentration of the intermediate C_B (pyruvate in this example) is zero and the steady-state rate $dC_C/dt = 0$. When activation at A occurs, pyruvate starts to build up to its steady-state concentration and the steady-state rate of the overall process is gradually approached. If one assumes that one is working with isolated independent enzyme reactions and intermediates which equilibrate freely with the medium, the transient time τ can be readily calculated. The intermediate (pyruvate) concentration will only reach a steady state if the ratio of the rates of formation and decomposition will keep it below the K_m value for lactate dehydrogenase.

The results of the experiments of Hess and Wurster indicate that in this case all the assumptions about the enzymes being structurally independent and the substrates being freely diffusible are justified. If the enzymes could interact in such a way that they could pass on their substrate the transient times would be modified. While there have been claims that there is some organization in concentrated solutions of glycolytic enzymes, a number of investigations of the type described here have failed to provide any evidence for kinetic interaction. This should not detract from the possibility that this type of investigation may give useful information on other systems. Curiously the application of these techniques to multi-enzyme systems known to be structurally organized has so far been neglected.

Some other applications of rapid recording techniques for observing whole enzyme systems are likely to provide kinetic evidence for the structure of the system or the point of perturbation. The question always is: how do the various parts of the system respond to the perturbation at some specific point? As we become more interested in the behaviour of some enzyme molecules of well-known structure and mechanism when they are incorporated in a complex mixture, we shall have to use these techniques in general biochemical work and not just for the study of reaction mechanisms (see Chance, 1967).

8.2. TRANSIENT KINETICS

8.2.1. Transients of Enzyme Intermediates

The observation of the transient formation and decomposition of different intermediates of the reaction of an enzyme with its substrate was initiated by Chance (for a survey see Chance, 1963) in his studies of peroxidase. Since that time transient changes in the absorption and fluorescence spectra of proteins, the prosthetic groups and the enzyme-bound substrates have all been used to follow the elementary steps of such reactions. In this way interesting information is obtained in addition to the kinetic data on the rates of inter-conversion of intermediates. The spectral changes themselves can provide information about the process which takes place during the transformation. For instance, free-radical formation can be identified, or changes in the polarity of the environment of the chromophore can give evidence for specific structures of transient intermediates. Changes in the ionization of intermediates can also be correlated with specific kinetic events with the use of indicators. Spectral changes of these and other kinds will be used to illustrate the reactions discussed. It should also be emphasized that in this way much more direct evidence for the existence of intermediates is obtained than from purely kinetic considerations. Intermediates are actually observed!

In this section we shall concern ourselves with studies of the spectral changes during the chemical transformation ES $\rightarrow$ EP. The formation of different intermediates of the enzyme with its substrates and products will be discussed in Section 8.2.3 under the title of partial reactions. Comment should be made about the fact that in some experiments the interconversion of enzyme substrate complexes are studied as part of the catalytic reaction $ES \rightarrow EP$.

An important task in the elucidation of a mechanism is to ascertain whether some identified intermediate X is on the direct pathway or on some side shunt:

$$E + S \rightleftharpoons ES \rightleftharpoons X \rightleftharpoons EP \rightleftharpoons E + P \qquad \text{Direct pathway}$$

$$E + S \rightleftharpoons ES \rightleftharpoons EP \rightleftharpoons E + P \qquad \text{Side shunt}$$
$$\underset{X}{\nwarrow \quad \nearrow}$$

The potentialities of this approach to the study of enzyme mechanisms have not yet been utilized very extensively. Present endeavours to make sensitive fluorescence and absorption spectroscopy, coupled to stopped-flow equipment, more readily available should contribute to their wider application. The interpretation of the perturbation of protein and reactant spectra during the formation of transient intermediates is probably the only method available for the elucidation of the structure of these intermediates. The classical studies of Chance with peroxidase and catalase provided the guidelines for both spectroscopic and kinetic analyses of the properties of transient intermediates (see Chance, 1963). With hydrolytic enzymes, which have been extensively investigated with rapid reaction techniques (Gutfreund, 1965), it has been possible to observe spectrally distinct enzyme–substrate compounds. So far, however, this method has not provided any evidence about the structure of intermediates in the chemical transformations catalysed by these enzymes.

The best examples to illustrate the interpretation of the observations of transient intermediates come from studies of flavoproteins. Palmer and Massey (1968) have presented a detailed discussion of the mechanisms of different flavoproteins and of the physical techniques used to study them. First we shall discuss two systems in which only the completely reduced prosthetic group is considered as an intermediate. For instance, in xanthine oxidase we wished to test the mechanism

$$\text{Enzyme-FAD} + \text{xanthine} \rightleftharpoons \text{Enzyme-FADH}_2 + \text{uric acid}$$

$$\text{Enzyme-FADH}_2 + O_2 \rightleftharpoons \text{Enzyme-FAD} + H_2O_2$$

The oxidized flavoprotein, like free flavine adenine dinucleotide, has a strong absorption band at 450 mμ. This band disappears when the enzyme is treated with xanthine under anaerobic conditions. Both the enzyme and free FAD are bleached by dithionite. The kinetic picture about to be presented here is somewhat oversimplified to illustrate the particular sequence of the cyclic oxidation-reduction of the enzyme. Xanthine oxidase is a complex enzyme which can undergo many reactions.

When xanthine oxidase is mixed with xanthine under anaerobic conditions the disappearance of the extinction at 450 mμ follows a first-order process. At saturating concentration of xanthine the first-order rate constant is 10·5 sec^{-1}. For our present simplification we assume that this first-order process represents the full reduction of all the enzymic FAD to FADH$_2$. In that case the following further deductions and correlations can be made from the observation of the overall reaction of the enzyme with saturating concentrations of xanthine and oxygen. When mixing the three components a rapid decrease in extinction at 450 mμ will be followed by a period of constant extinction. Finally, the enzyme will be completely bleached if xanthine is in excess or it will regain its full extinction at 450 mμ if oxygen is present in excess:

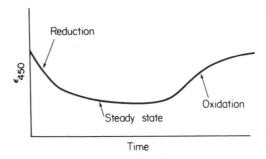

If the reaction cycle simply goes through fully reduced enzyme molecules (by the step characterized by $k = 10·5 \text{ sec}^{-1}$) then the rate of reduction to the steady state must occur with an apparent rate constant which is the sum $(k_2 + k_4)$ of the rate constants for the steps:

$$\text{Xanthine-Enzyme-FAD} \overset{k_2}{\rightleftharpoons} \text{Uric acid-Enzyme-FADH}_2$$

and

$$\text{Oxygen-Enzyme-FADH}_2 \overset{k_4}{\rightleftharpoons} \text{H}_2\text{O}_2\text{-Enzyme-FAD}$$

If the rate constant for the re-oxidation step is calculated from the above information we have two further tests for the consistency of this simple mechanism. First the turnover rate of the enzyme must be

$$k_{\text{cat}} = \frac{k_2 k_4}{k_2 + k_4}$$

and secondly the ratio of reduced to oxidized enzyme in the steady state must be k_2/k_4 (see Gutfreund and Sturtevant, 1959).

An enzyme which appears to go through such a simple cycle and which has been investigated in greater detail is glucose oxidase from *Aspergillus niger*. Gibson and his colleagues have illustrated the use of analog computer representations of proposed mechanisms to provide evidence for the correct interpretation of kinetic results. The comparison between experimental data and computer-simulated mechanism, which is illustrated in Figure 45, was

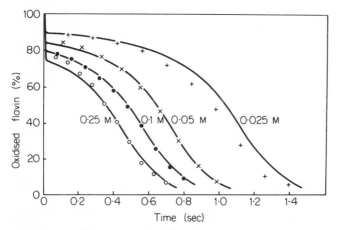

Figure 45 The sets of points represent extinction changes calculated from stopped-flow experiments. At zero time solution A (containing 1.76×10^{-5} glucose oxidase, $2.6 \times 10^{-4}\,M\,O_2$) was mixed with solution B (containing glucose-1-^{2}H to give the final concentrations indicated on the curves). The reaction mixtures also contained $0.1\,M$ potassium acetate pH 5.5 and $10^{-3}\,M$ KCN. The three phasic reduction of the flavine group of the enzyme is due to (1) a rapid reduction to the steady-state level, (2) a slowly declining steady state and (3) complete reduction on anaerobiosis. The solid lines are analog computer simulations of the process using the simple model and rate constants determined by the method discussed for xanthine oxidase on p.191 (Bright and Gibson, 1967)

part of a study of the deuterium isotope effect on the reduction of the enzyme by substrate. A large decrease in the rate constant Glucose-Enzyme-FAD $\rightarrow$ Gluconolactone-E-FADH$_2$ was found when 1-deuterated glucose was used as a substrate.

In a number of flavoprotein systems two forms of the reduced enzyme can be produced: E-FADH$_2$ and the free radical E-FADH$\cdot$. The key question is whether the kinetic behaviour of the free radical is consistent with it being a compulsory intermediate in the normal enzymic turnover. In some cases the fully reduced enzyme-$FADH_2$ can be excluded from the normal pathway. For instance, excess substrate under anaerobic conditions will reduce L-amino acid oxidase completely. It is found that the oxidation of this form of the enzyme is slower than the enzymic turnover. This indicates that the fully reduced form cannot be a normal reaction intermediate.

The first example of a thorough kinetic investigation into the role of a semiquinone intermediate is Massey, Gibson and Veeger's (1960) study of lipoyl dehydrogenase. Chemical studies on this enzyme provided evidence that reduction of this flavoprotein by dihydrolipoic acid or NADH results in the reduction of an S—S bridge and the interaction of the —SH groups thus formed with FADH·. The flavine free radical has a typical absorption band in the region 500–600 mμ. The kinetic investigations were designed to test the role of the semiquinone as an obligatory intermediate in the same way as the role of the completely reduced E-$FADH_2$ was investigated in the above examples.

The steady state and transient kinetic data can be compared to test the following scheme:

$$E(FAD)_2 + AH_2 \; \underset{}{\overset{k_1}{\rightleftharpoons}} \; AH_2E(FAD)_2 \; \underset{}{\overset{k_2}{\rightleftharpoons}} \; E(FADH \cdot)_2 + A$$

$$E(FADH \cdot)_2 + B \; \underset{}{\overset{k_3}{\rightleftharpoons}} \; BE(FADH \cdot)_2 \; \underset{}{\overset{k_4}{\rightleftharpoons}} \; E(FAD)_2 + BH_2$$

The rate constants k_2 and k_4 can be determined independently at substrate saturation from stopped-flow observations at 530 mμ. If the intermediate observed at 530 mμ, $E(FADH \cdot)_2$, is on the direct pathway of the reaction then

$$K_m \text{ (for } AH_2) = [k_4(k_{-1} + k_2)]/[k_1(k_2 + k_4)]$$

$$K_m \text{ (for } B) = [k_2(k_{-3} + k_4)]/[k_3(k_2 + k_4)]$$

and

$$V_{max}/C_E^\circ = k_{cat} = k_2 k_4/(k_2 + k_4)$$

8.2.2. Transients of Reactants and Single Turnovers

In this section kinetic equations will be derived for the rate of appearance of product during the transient period between mixing enzyme and substrate and the time when the steady-state rate is reached. These equations can then be used to analyse reaction records in terms of numbers of intermediates and rate constants. The rate of formation of one product at the time will be considered and the situation simplified in such a way that when we are not dealing with a single substrate system all but one of the substrates will be present at saturating concentrations.

The most complicated reaction scheme which we shall consider

$$\text{E} + \text{S} \underset{}{\overset{k_1}{\rightleftharpoons}} \text{ES} \underset{}{\overset{k_2}{\rightleftharpoons}} \text{E*S} \underset{}{\overset{k_3}{\rightleftharpoons}} \text{E*P} \underset{}{\overset{k_4}{\rightleftharpoons}} \text{EP} \underset{}{\overset{k_5}{\rightleftharpoons}} \text{E} + \text{P}$$

is in fact the most complex process for which experimental evidence will be provided under any one condition. The amount of detailed information which can be obtained depends on the position of the slowest step among the first-order interconversions (k_2 to k_5). The conditions of substrate concentrations imposed above result in only one second-order step in each direction having to be considered.

Let us first derive the equation for the approach to the steady state for a system in which only one intermediate, ES, occurs at a significant level in the steady state. In such a system the steady-state velocity $v = k_2 C_{ES} = dC_P/dt$. At zero time when enzyme and substrate are mixed $C_{ES} = 0$ and $dC_P/dt = 0$. As C_{ES} is formed dC_P/dt increases and d^2C_P/dt^2 is a measure of the rate of formation of ES:

$$dC_{ES}/dt = k_1 C_S(C_E^\circ - C_{ES}) - (k_{-1} + k_2)C_{ES}$$

if dC_{ES}/dt is expressed as $(d^2C_P/dt^2)k_2$ and C_{ES} as $(dC_P/dt)/k_2$ we obtain

$$d^2C_P/dt^2 + dC_P/dt(k_1 C_S + k_{-1} + k_2) = k_1 k_2 C_S C_E^\circ \qquad (8.4)$$

This equation can only be solved analytically under the condition that C_S is sufficiently large to be regarded as constant. If we write equation (8.4) in the form

$$d^2C_P/dt^2 + (dC_P/dt)A = B \qquad (8.5)$$

this can be integrated to give

$$C_P = \frac{B}{A}t + C e^{-At} + D \qquad (8.6)$$

In equation (8.6) C and D are integration constants which can be evaluated from the boundary conditions at $t = 0$, $C_P = 0$, $C_{ES} = 0$ as follows:

$$dC_P/dt = 0 = \frac{B}{A} - AC$$

Therefore,

$$C = B/A^2$$

$$D = C_P^\circ \,(\text{at } t = 0) - C = C_P^\circ - B/A^2$$

$$C_P = \frac{B}{A}t + \frac{B}{A^2}(e^{-At} - 1)$$

$$C_P = \frac{k_1 k_2 C_S C_E^\circ}{k_1 C_S + k_{-1} + k_2}t + \frac{k_1 k_2 C_S C_E^\circ}{(k_1 C_S + k_{-1} + k_2)^2}e^{-[k_1 C_S + k_{-1} + k_2]t}$$

$$+ \frac{k_1 k_2 C_S C_E^\circ}{(k_1 C_S + k_{-1} + k_2)^2} + C_P^\circ \qquad (8.7)$$

The physical meaning of the three terms of this equation can be illustrated diagrammatically as follows:

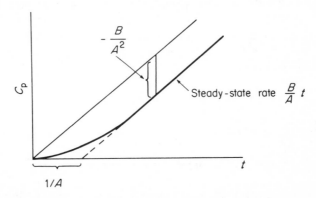

For this particular case, with k_2 rate-limiting, $K_m = (k_2 + k_{-1})/k_1$ and the intercept $1/A$ is $1/k_1(C_S + K_m)$. In theory it is, therefore, possible to evaluate k_1 and it is also possible to ascertain, from the characteristic dependence on C_S, whether the intercept does represent the acceleration due to this second-order process. In practice it turns out that all attempts to determine the rate of the first step of enzyme–substrate combination resulted in values $k_1 \geqslant 5 \times 10^6 \,\text{M}^{-1}\,\text{sec}^{-1}$. It would be difficult to carry out such experiments with C_S much smaller than $10^{-4} \,\text{M}$ and the intercept of milliseconds cannot be determined quantitatively. If it is not possible to use one of the more direct methods for determining k_1 discussed in the following sections, the observation of transients can give useful estimates of lower limits for this rate constant.

If a transient acceleration of product formation is observed, which does not disappear as C_S is increased, then this is due to a first-order process involved in the approach to the steady state of a second enzyme–substrate compound (for instance, $ES \xrightarrow{k_2} E^*S$). In such a case the intercept of the steady-state line extrapolated back to the time axis is the reciprocal of the apparent rate constant for the approach to the steady state of C_{E^*S}. In the simplest case, when k_{-2} and k_{-3} can be neglected and k_3 is rate limiting, the intercept is $1/(k_2 + k_3)$ if $C_S \gg (k_2 + k_{-1})/k_1$.

While the acceleration of product formation has limited application for the direct evaluation of rate constants, it is important for this aspect of transient kinetics to be properly understood before the more informative next phase can be analysed. In Figure 42 the rate of product formation shows three distinct phases:

(1) Acceleration due to approach to the steady-state concentration of all enzyme intermediates prior to product formation.
(2) Rapid formation of enzyme–product intermediates to steady-state level.
(3) The steady-state rate of product formation.

In practice the interpretation of transients of reactants has been studied in a complementary manner in two different types of system. In the case of a large number of hydrolytic enzymes the reaction can be described by

$$E + AB \underset{}{\overset{k_1}{\rightleftharpoons}} EAB \underset{}{\overset{k_2}{\rightleftharpoons}} EA + B$$

$$EA \xrightarrow{k_3} E + A$$

The steps characterized by k_2 and k_3 are sometimes controlled by a previous isomerization, for instance $EAB \rightleftharpoons E^*AB \rightarrow$ or $E^*A \rightleftharpoons EA \rightarrow$. The absence of product allows one to write some steps in an irreversible form. In the case of alkaline phosphatase for instance, phosphate esters with chromophoric leaving groups B can be used to study the reaction in the transient phase. The rate of appearance of B is a measure of the rate of appearance of EA plus A. The rate of interconversion of an enzyme–substrate compound to an enzyme–product compound as well as the steady-state liberation of product can be calculated from experimental records of the type shown in Figure 46. Before deriving the necessary equations, another type of experiment will be described which gives distinct information about the transformation $ES \rightarrow EP$ and the steady rate of formation of product. In this case there is a change in spectrum during the reaction $S \rightarrow P$ and some suitable wavelength is chosen for the observation, at which the extinction of the enzyme-bound forms of the product is identical with that of the free product. An example of this type, investigated in some detail, is the reduction of enzyme-bound NAD to NADH catalysed by horse liver alcohol dehydrogenase:

$$E^{NAD}_{Ethanol} \rightleftharpoons E^{NADH}_{Aldehyde} \rightarrow E^{NADH} \rightarrow E + NADH$$

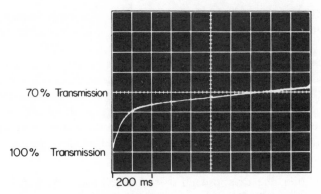

Figure 46   Record of a stopped-flow experiment: the final reaction mixture contained 8·7 μM alkaline phosphatase (*E. coli*) and 0·67 mM 2,4-dinitrophenyl phosphate in 0·1 M acetate pH 5·5. The rapid transient formation of dinitrophenol, corresponding to one equivalent of the dimeric enzyme present, is followed by the steady-state reaction. The progress of the reaction was followed at 360 mμ

Detailed kinetic analysis also produced evidence for isomerization steps of the type $E^{NAD}_{Ethanol} \rightleftharpoons E^{*NAD}_{Ethanol}$ and $E^{*NADH}_{Aldehyde} \rightleftharpoons E^{NADH}_{Aldehyde}$ (Shore and Gutfreund, 1970). Second-order steps in the forward direction are neglected on the assumption that C_{NAD} and $C_{Ethanol}$ are very high and second-order reversal steps are neglected on the assumption that C_{NADH} and $C_{Aldehyde} \approx 0$. In the case of the record shown in Figure 47 of an experiment with horse

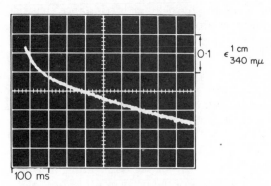

Figure 47   The initial transient of the reaction of horse liver alcohol dehydrogenase (30 μM in sites) with deuterated ethanol (50 mM) and NAD$^+$ (20 mM) at pH 8. After the rapid production of 1 mole of enzyme-bound NADH follows the steady-state rate of NADH formation

liver alcohol dehydrogenase it is again possible to calculate the rate of inter-conversion of enzyme-bound substrate to enzyme-bound product, as well as the steady-state rate of product formation.

An interesting feature of the two experiments shown above is that in one case one can show that the transformation of enzyme–substrate complex to the enzyme–product complex is controlled by an isomerization of the enzyme–substrate complex, prior to the chemical reaction (Trentham and Gutfreund, 1968), while in the other case the observed rate of transformation is controlled by the chemical process—hydride transfer (Shore and Gutfreund, 1970). In the case of alkaline phosphatase the experiment shown in Figure 46 can be repeated with a series of phosphate esters with leaving groups with widely differing pKs. It is found that the rate of the reaction $EAB \rightarrow EA + B$ is quite independent of the nature of B. This indicates that in the process

$$EAB \rightarrow E^*AB \rightarrow E^*A + B$$

the first step, an isomerization of the enzyme–substrate complex, is rate limiting. Evidence for this type of mechanism has been summarized by Gutfreund (1971).

When the experiment with liver alcohol dehydrogenase is carried out in the presence of deuteroethanol, it is found that the transformation

$$E^{NAD^+}_{Ethanol} \rightarrow E^{NADH}_{Aldehyde} + H^+$$

is about 1/6th of the rate compared with C_2H_5OH as substrate. This indicates that the hydride transfer step is being observed and that the use of a wide range of alcohols will provide valuable correlation between structure and reactivity. With pig heart lactate dehydrogenase the transient rate of the reaction $E^{NAD^+}_{Lact} \rightarrow E^{NADH}_{Pyruvate}$ is not affected by deuterium substitution in lactate. The conformation change must determine the rate of the reaction.

The derivation of the rate equation for the transient appearance of product, enzyme bound and free, will now be given in terms of the model $E + AB \overset{k_1}{\underset{}{\rightleftharpoons}} EAB \overset{k_2}{\rightarrow} EA + B, EA \overset{k_3}{\rightarrow} E + A$. It is assumed that C_{AB} is large enough so that, on the time scale considered, the first transient phase (the equilibration $E + AB \rightleftharpoons EAB$) is sufficiently rapid to be neglected and the total enzyme concentration $C^\circ_E = C_{EAB} + C_{EA}$ and C_E is negligible. The rate of formation of EA is given by

$$dC_{EA}/dt = k_2 C_{EAB} - k_3 C_{EA}$$

$$= k_2 C^\circ_E - (k_2 + k_3) C_{EA}$$

$$dC_{EA} \Big/ \left(\frac{k_2 C^\circ_E}{k_2 + k_3} - C_{EA} \right) = (k_2 + k_3)\, dt \qquad (8.8)$$

integration and introduction of the boundary condition that, at $t = 0$, $C_{EA} = 0$ gives

$$C_{EA} = \frac{k_2 C_E^\circ}{k_2 + k_3}[1 - e^{-(k_2 + k_3)t}] \tag{8.9}$$

If we now introduce $C_{EAB} = C_E^\circ - C_{EA}$ and $dC_B/dt = k_2 C_{EAB}$, then

$$C_{EAB} = C_E^\circ - \frac{k_2 C_E^\circ}{k_2 + k_3}[1 - e^{-(k_2 + k_3)t}] \tag{8.10}$$

and

$$dC_B/dt = \frac{k_2 C_E^\circ}{k_2 + k_3}[k_3 + k_2 e^{-(k_2 + k_3)t}] \tag{8.11}$$

this can be integrated to give

$$C_B = \frac{k_2 k_3}{k_2 + k_3}C_E^\circ t - \frac{k_2}{k_2 + k_3}C_E^\circ e^{-(k_2 + k_3)t} + \text{constant}$$

The constant of integration can be evaluated from the boundary condition that, at $t = 0$, $C_B = 0$ and

$$\text{Constant} = [k_2/(k_2 + k_3)]^2$$

Therefore,

$$C_B = C_E^\circ \left\{ \frac{k_2 k_3}{k_2 + k_3}t + \left(\frac{k_2}{k_2 + k_3}\right)^2 [1 - e^{-(k_2 + k_3)t}] \right\} \tag{8.12}$$

The terms of this equation are illustrated in the following diagram:

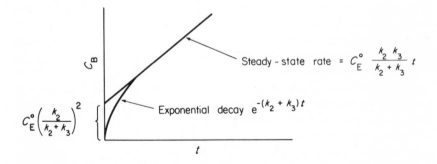

If the experimental observation provides a record of the sum of free and enzyme-bound product, as in the example of Figure 47, the concentration of total product is substituted for C_B in the above equations. Continuing with the assumption that only one enzyme–product compound is present at a significant concentration, the rate constant for the formation of this compound can be evaluated as follows. The extended Guggenheim method discussed in Section 6.1.2 can be applied in two ways to the calculation of the rate of approach to the steady state, i.e. to the rate of formation of the steady-state concentration of the enzyme–product intermediate. The fully arithmetical and the partially graphical procedure are illustrated in Figure 48 on an example in which k_2 and k_3 are of similar magnitude. The rate of approach to the steady state is controlled by $k_2 + k_3$. When $k_2 \gg k_3$ equation (8.12) becomes much simpler and its terms can be interpreted without additional information. In some systems it is possible to inhibit the rate of decomposition of the enzyme–product complex without affecting its formation, which is one way of making k_2 much larger than k_3.

One particular ambiguity in the interpretation of the transient kinetics of enzyme–product compound formation needs to be emphasized. The intercept $C_E^\circ[k_2/(k_2 + k_3)]^2$, which is obtained from extrapolating the steady-state velocity to $t = 0$, is often used to check either the correct enzyme site concentration or the ratio k_2/k_3. It is becoming apparent, however, that the sites on polymeric enzymes are not necessarily equivalent and in some cases alternate in their activity. Chromophoric substrates have been widely used to determine the number of simultaneously available active sites. In the case of E. coli phosphatase this kind of analysis shows that only one site of the dimer of identical units is available at the time. It is also possible to evaluate the number of sites (n) active at any one time from

$$\text{Turnover number} = n\frac{k_2 k_3}{k_2 + k_3}$$

If we remove the restriction that only one enzyme–product complex occurs at a significant concentration, then we must expect that the transient appearance of enzyme-bound product will have to be described by a sum of exponentials instead of the single term $e^{-(k_2+k_3)t}$. In practice no case has yet been found in which the transients of product formation could contribute to the other available evidence for a second enzyme–product compound. General equations for the transients of reactants in multistep processes have been derived by Darvey (1968). The most profitable approach to the determination of the complete mechanism of an enzyme reaction is the use of all the methods (kinetic, spectral, isotopic) to formulate a model. The computer simulation of this model should fit both the transients of reactants and the transients of intermediates. Although a good fit of computed to experimental curves provides no proof for a mechanism, the results of transient kinetics may put a severe limitation on alternatives.

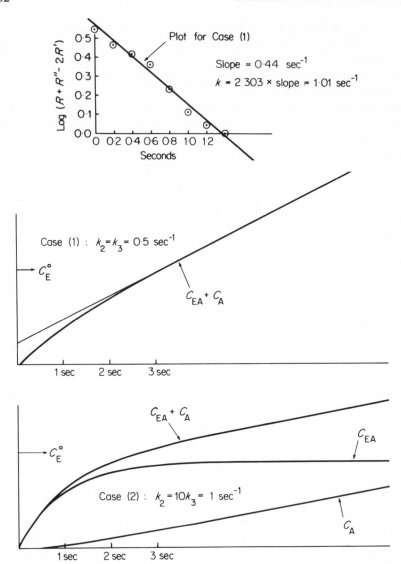

Figure 48 The computer simulation of two cases of transient product formation in the reaction

$$E + AB \rightleftharpoons EAB \xrightarrow{k_2} EA + B$$

$$EA \xrightarrow{k_3} E + A$$

In both cases the equilibration $E + AB \rightleftharpoons EAB$ is assumed to be rapid compared with the rest of the reaction and free C_A and C_B to be sufficiently small for the second and third steps to be considered irreversible.

8.2.3. Partial Reactions of Enzymes with Substrates

In this section we shall concern ourselves with some of the methods which give information about the interaction between enzymes and substrates prior to any chemical transformations. One of the methods used for such studies, the chemical relaxation technique, is discussed separately in Section 8.3.1. Three different types of kinetic investigations with rapid flow techniques can be used:

(1) Reactions of a multi-substrate enzyme with incomplete sets of substrates.
(2) Reactions with substrate analogs (competitive inhibitors).
(3) Single turnover reactions with evidence that the rate-limiting step occurs prior to any chemical transformation.

The second-order rate constants of the formation of initial complexes of enzymes with their substrates can only be determined by flow methods if two conditions hold. First, one must have a very sensitive method for monitoring the change from free to enzyme-bound substrate. Secondly, the dissociation constant of the complex must be small so that a significant amount is formed at low enzyme and substrate concentrations. The reason for these conditions is simply that most of the second-order rate constants for these processes are between 10^7 and 10^8 M^{-1} sec^{-1}. For good results the first halftime of a flow measurement should not be less than 5 msec. For a second-order reaction

$$t_{\frac{1}{2}} = 1/kC_S \text{ (when enzyme and substrate concentration } C_S \text{ are equal)}$$

$$5 \times 10^{-3} = 1/(5 \times 10^7 \times C_S)$$

$$C_S = 4 \, \mu M$$

This simple calculation shows that one has to be able to carry out the measurements at enzyme and substrate concentrations in the range of $1–10 \, \mu M$.

Figure 48 (*continued*)
For case (1), when $k_2 = k_3 = 0.5 \, sec^{-1}$, the extended Guggenheim procedure is demonstrated. The readings R of the recorded concentration $(C_{EA} + C_A)$ are taken at time t, while readings R' and R'' are taken at times $t + 2 \, sec$ and $t + 4 \, sec$, respectively. The plot of $\log (R + R'' - 2R')$ against time gives (after conversion to natural logarithms) the rate constant for the approach to the steady-state concentration of EA. This rate constant is the sum $k_2 + k_3$. The scatter of points is due to inaccuracies of reading the small changes of R from the record.
For case (2) $k_2 = 10k_3 = 1 \, sec^{-1}$.

The only monitoring technique which provides adequate sensitivity to follow a reaction with such low reactant concentrations is the measurement of fluorescence changes. The reactions of dehydrogenases with NADH have been studied in a number of laboratories (for a survey see Gutfreund, 1971). In many systems NADH fluorescence is altered considerably on enzyme–complex formation as well as during subsequent ternary complex formation with the second substrate. Furthermore, NADH binds very tightly to most dehydrogenases ($K \approx 1\,\mu\text{M}$ dissociation constant) and an appreciable proportion of the reactants end up in the form of the complex.

One of the aims of kinetic investigations of substrate binding is to establish whether this process occurs in one single second-order step or whether the initial contact is followed by a first-order isomerization. We are first concerned with the correct determination of the second-order rate constant and shall subsequently discuss the isomerization phenomena. Visual inspection of Figure 49 indicates an apparent first-order reaction between NADH and

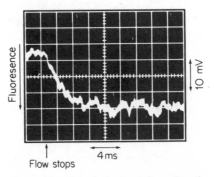

Figure 49 Association reaction after the mixing of M_4 lactate dehydrogenase (8 μM-enzyme sites) with 8 μM NADH in the stopped-flow fluorimeter. The pH was 7·2 and the temperature 22° C (from Stinson and Gutfreund, 1971)

lactate dehydrogenase to form an E-NADH complex. However, this can be compared with the computer simulations when 62 per cent. of the reactants are in the complex form at equilibrium (Figure 22). A logarithmic plot of the computer data shows them to be indistinguishable from a first-order rate process. A second-order plot will only approach to give the right answer for the second-order constant when the reaction goes to over 90 per cent. completion. For experiments like that illustrated in Figure 49 correct evaluation of the rate constants requires the use of the equations for reversible reactions given in Section 6.2.1 (equation (6.14)).

Another comment of caution for the evaluation of second-order rate constants should be directed at the necessity for the determination of the correct endpoint of the reaction. In the case of a true second-order record the reaction becomes very slow towards the end and it becomes much more difficult than for a first-order process to separate instrumental or chemical drift.

The critical test for the order of a reaction comes from experiments over a range of initial reactant concentrations. Some mechanisms can only be properly explored if both signal and time resolution permit the measurements to be carried out over a wide range of substrate concentration. A mechanism of special interest is the two-step binding process

$$E + S \underset{}{\overset{k_1}{\rightleftharpoons}} ES \underset{}{\overset{k_2}{\rightleftharpoons}} E^*S$$

Let us assume for the moment that the observed signal is due to the appearance of E*S. If the rate of equilibration of ES is much faster than of E*S, at the enzyme and substrate concentration used, a first-order process with rate constant

$$k = k_{-2} + k_2/(1 + K_1/C_S) \qquad \begin{pmatrix} K_1 = k_{-1}/k_1 \\ K_2 = k_{-2}/k_2 \end{pmatrix} \qquad (8.13)$$

is observed. The amplitude of the reaction will increase with increasing substrate concentration until $C_S \gg K_1 K_2$. This derivation is widely used for the kinetics of ligand binding and also for the determination of rate constants for chemical steps subsequent to a rapid initial substrate binding process. This treatment should be compared with the relaxation of such a system (see p. 213).

If two distinct signals are observed for the formation of ES and E*S additional information can be obtained about the nature of the complexes. It is also possible, in such cases, to correlate K_1 calculated from the dependence of the reaction amplitude on C_S, with K_1 derived from a plot of $1/C_S$ against $1/k$ (see equation (8.13)).

There are a number of methods available for the study of the rate of dissociation of substrates or other ligands from their enzyme complexes. Chemical relaxation studies of dissociation processes are discussed in detail in Section 8.3.1, but some of the relevant experiments in flow equipment require similar interpretations. Concentration jump (dilution), pH jump and ionic strength changes can be achieved by mixing suitable solutions in a stopped-flow apparatus. If these changes result in small perturbation of equilibria they can be interpreted in terms of relaxation spectra. Large perturbations need to be interpreted with caution.

A method which has been used successfully for the measurement of the rate of dissociation of NADH from a number of its complexes with dehydrogenases, is the displacement method. Let us go from the particular to the general and first illustrate the procedure with a particular example. NADH has an increased fluorescence intensity (and blue shift) when bound to heart or skeletal muscle lactate dehydrogenase. The dissociation E-NADH → E + NADH can be observed as a decrease in fluorescence intensity. The dissociation constant of the skeletal muscle enzyme–NADH complex is about $2\,\mu\text{M}$. Consequently in a solution containing $20\,\mu\text{M}$ enzyme sites and $20\,\mu\text{M}$ NADH a high proportion of the binding sites are occupied. If this solution of E-NADH is mixed with a solution containing a high concentration of some compound X which competes for the NADH binding site on the enzyme, the following reactions take place:

$$\text{E-NADH} \underset{}{\overset{k_{-1}}{\rightleftharpoons}} \text{E} + \text{NADH}$$

$$\text{E} + \text{X} \underset{k_2}{\rightarrow} \text{EX}$$

Figure 50 shows the record of an experiment in which X, the displacement reagent, is NAD^+. If $k_{-1} \ll k_2 C_{\text{NAD}^+}$ and $k_2 C_{\text{NAD}^+} \gg k_1 C_{\text{NADH}}$ the signal due to the disappearance of E-NADH can be interpreted in terms of the rate constant k_{-1} for the dissociation of this complex.

The study of displacement reactions should always be carried out with a number of different displacement reagents and, if possible, it should be complemented with other dissociation rate measurements. There are a

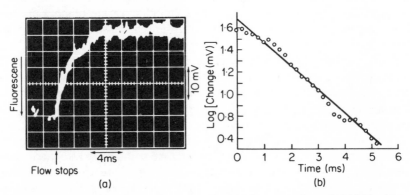

Figure 50   (a) Displacement of M_4 lactate dehydrogenase-bound NADH with NAD^+. Enzyme-NADH ($20\,\mu\text{M}$) was mixed with an equal volume of $30\,\text{mM-NAD}^+$ in the stopped-flow fluorimeter. The pH was 7·2 and the temperature 22° C. (The traces of two identical experiments are superimposed in (b) First-order Guggenheim plot of the reaction shown in (a). (From Stinson and Gutfreund, 1971)

number of interesting effects which can be discovered by comparing different methods. For instance, in the case of dimeric phosphatase from *E. coli* the two intrinsically identical binding sites cannot both bind substrate (or its analog) at the same time in the same way

$$SE\,E^*S \rightleftharpoons SE^*\,ES$$

In such a case displacement from one site depends on the occupation of the other.

Another type of situation in which different methods can give different results for dissociation rate constants can be illustrated with some experiments on haemoglobin. In Section 4.3.3 it was pointed out that the affinity of haemoglobin molecules for oxygen and similar ligands depends on the number of sites occupied on the tetrameric molecule. There are a variety of ways in which the rate of dissociation of oxygen from haemoglobin can be determined experimentally (dilution, relaxation, displacement and chemical removal of O_2). If we compare results obtained from

(1) $Hb(O_2)_4 \rightleftharpoons Hb + 4O_2$
$\qquad\quad \downarrow {+4CO}$
$\qquad Hb(CO)_4$

and

(2) $Hb(O_2)_4 \rightleftharpoons Hb + 4O_2$ (oxygen rapidly removed by excess dithionite)

In the first case the rate constant obtained is that for the dissociation of fully liganded haemoglobin, because all sites will be occupied all the time either by O_2 or CO. In the second case the rate constant will change as the reaction proceeds and as successive oxygen molecules dissociate from the haemoglobin molecules.

If a polymeric protein molecule has intrinsically different (non-interacting) binding sites, it is of interest to find out whether the difference in affinity of these sites is due to differences in association rates or dissociation rates. An example of a system in which one would wish to determine the rate of dissociation from different sites are hybrids of two forms of an enzyme. In heart lactate dehydrogenase (H_4LDH) and skeletal muscle lactate dehydrogenase (M_4LDH) the four sites of each homogeneous tetramer are identical and independent. However, the rate of dissociation of NADH from its complex with the H_4 enzyme is slower than the rate of dissociation from the complex with M_4LDH. The difference in the two rates depends on pH and other ions in the solution. Five hybrids of H_4 and M_4LDH can be made and it is of interest to see whether the individual sites of, say, H_2M_2LDH behave as they do in their respective homogeneous tetramer. If the rate of dissociation is measured by the displacement method one obtains a record of two simultaneous first-order reactions. To be able to resolve the records of superimposed first-order reactions the rate constants must differ by at least a factor of three if the amplitudes of the components are about equal.

The method for resolving the two rate constants for parallel first-order processes is illustrated in Figure 24. In the case of the dissociation of NADH from the H_2M_2LDH hybrid the two rate constants obtained are the same as those found for dissociation from H_4 and M_4LDH, respectively. The amplitudes for the two components were not quite equal. This could be due partly to the enzyme preparation not containing exactly 50 per cent. each of the heart and skeletal muscle units. The other uncertainty is the relative fluorescence changes on NADH dissociation from the M and the H units.

There is another type of kinetic experiment with enzymes which can be classified as the study of partial reactions. In many cases strong binding of one substrate makes it possible to prepare a solution of a more or less stoichiometric compound between an enzyme and that substrate. A particular step of the reaction of this compound with the second substrate can often be studied in isolation. For instance, glyceraldehyde 3-phosphate dehydrogenase forms an acyl enzyme compound with 1,3-diphosphoglycerate and the reaction of this intermediate with NADH can be studied stoichiometrically. For another example we again make use of the small dissociation constant of the complex lactate dehydrogenase-NADH. Figure 51 shows that the oxidation of NADH bound to the four sites of lactate dehydrogenase proceeds as a single first-order process, indicating that the sites are identical and independent. Some other interesting information can be obtained from the study of the spectral change which occurs during the reaction

$$E^{NADH} + \text{pyruvate} + H^+ \rightarrow E^{NAD^+}_{\text{lactate}}$$

First it is possible to study the saturation curve for pyruvate. If the apparent K_m calculated from this is identical with the K_m obtained from steady-state measurements then the steps occurring after the oxidation of enzyme bound NADH cannot be rate limiting (see Stinson and Gutfreund, 1971).

It can be shown that the first-order reaction observed for the step $E^{NADH}_{pyr} \rightarrow E^{NAD^+}_{lact}$ (at pyruvate saturation) is not limited by the chemical step of hydride transfer. The reaction proceeds at the same rate when stereospecifically deuterated NADD is substituted for NADH. Stinson and Gutfreund (1971) have proposed that an isomerization of the ternary complex $E^{NADH}_{pyruvate}$, prior to the hydride transfer process, is rate-limiting for this partial reaction.

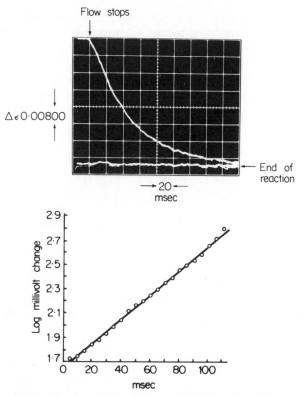

Figure 51   A record of the oxidation of NADH
(10 μM) bound to pig M_4 lactate dehydrogenase
(14 μM in active sites) by 1 mM pyruvate in 0·1 M
glycine–NaoH buffer at pH 9. The first-order plot of
the time course of the reaction is also given. The reaction
was carried out at 22° and the oxidation of NADH was
observed in a stopped-flow apparatus at 340 mμ (from
Stinson and Gutfreund, 1971)

8.3. RELAXATION KINETICS

8.3.1. Enzyme Substrate Combination

For the purpose of this discussion the term enzyme substrate combination
is defined as the binding of any substrate or substrate analog which can be
studied at equilibrium $E + S \rightleftharpoons ES$, without further reaction. This includes
one substrate partner of a two-substrate enzyme, or two substrates in a
three-substrate enzyme, substrates with or without metal ligands and
modified substrates which will simulate the binding process only. The

principle of relaxation kinetics will be discussed in terms of examples taken from temperature jump experiments, but the general equations developed are applicable to the interpretation of all single step-function perturbations of systems at equilibrium.

When a system at equilibrium is suddenly heated the concentrations C_E, C_{ES}, C_S are no longer equilibrium concentrations. The equilibrium concentrations for the new condition $\bar{C}_E$, $\bar{C}_{ES}$, $\bar{C}_S$, are to be reached through relaxation of the process $E + S \underset{k_1}{\overset{k_1}{\rightleftharpoons}} ES$ to its new equilibrium:

$$\begin{array}{ll} \rule{5cm}{0.4pt} & C^\circ_{ES} \\ \rule{5cm}{0.4pt} & C_{ES} \\ \}\Delta C_{ES} \rule{3cm}{0.4pt} & \bar{C}_{ES} \end{array}$$

Applying normal kinetic procedures one can write for the change in C_{ES}

$$\frac{dC_{ES}}{dt} = k_1 C_E C_S - k_{-1} C_{ES}$$

as indicated on the above diagram

$$\frac{d(\bar{C}_{ES} + \Delta C_{ES})}{dt} = k_1(\bar{C}_E + \Delta C_E)(\bar{C}_S + \Delta C_S) - k_{-1}(\bar{C}_{ES} + \Delta C_{ES})$$

$$\frac{d\bar{C}_{ES}}{dt} + \frac{d\Delta C_{ES}}{dt} = k_1\bar{C}_E\bar{C}_S + k_1(\bar{C}_S\Delta C_E + \bar{C}_E\Delta C_S) + k_1\Delta C_E\Delta C_S$$
$$- k_{-1}\bar{C}_{ES} - k_{-1}\Delta C_{ES}$$

from the equilibrium condition it is clear that

$$\frac{d\bar{C}_{ES}}{dt} = k_1\bar{C}_E\bar{C}_S - k_{-1}\bar{C}_{ES} = 0$$

and it follows that

$$\frac{d\Delta C_{ES}}{dt} = k_1(\bar{C}_S\Delta C_E + \bar{C}_E\Delta C_S) + k_1\Delta C_E\Delta C_S - k_{-1}\Delta C_{ES}$$

The all-important condition for the correct interpretation of chemical relaxations is that the perturbation in concentrations must be small compared with the total concentrations. This is defined algebraically by ΔC_E, $\Delta C_S \ll \bar{C}_E$, $\bar{C}_S$, which are the conditions for linearizing the differential equations. Use of this condition results in

$$\frac{d\Delta C_{ES}}{dt} = k_1(\bar{C}_S\Delta C_E + \bar{C}_E\Delta C_S) - k_{-1}\Delta C_{ES}$$

If we now add the following conditions on the concentration parameters,

$$\Delta C_E = \Delta C_S \quad \text{and} \quad \Delta C_S + \Delta C_{ES} = 0 \ (C_S + C_{ES} = \text{constant})$$

we obtain the following equation

$$\frac{d\Delta C_{ES}}{dt} = -[k_1(\bar{C}_E + \bar{C}_S) + k_{-1}]\Delta C_{ES}$$

$$\Delta C_{ES} = \Delta C_{ES}^{\circ} \, e^{-[k_1(\bar{C}_E + \bar{C}_S) + k_{-1}]t} = \Delta C_{ES}^{\circ} \, e^{-1/\tau} \tag{8.14}$$

The relaxation time τ of a system is defined as the time taken to decay to $1/e$ of the value at zero time. The relaxation time of the simple association–dissociation process used for our first example is dependent on the forward and reverse rate constants and the concentrations $\bar{C}_E + \bar{C}_S$.

The rate constants k_1 and k_{-1} for such an equilibrium process can be readily evaluated from relaxation experiments carried out over a range of concentrations. As shown in Figure 52 a plot of $C_E + C_S$ against $1/\tau$ results in a straight line with the intercept (the concentration-independent term) equal to k_{-1} and the slope equal to k_1. A straight line indicates that these are the only rate processes taking place on this time scale. It does not exclude other more rapid relaxations.

After the above complete treatment of one practical but simple example, some quite general comments about the theory of relaxations should be made. Afterwards a number of other examples will be treated which are known to have application to enzyme–substrate interactions. The term relaxation is applied to a variety of phenomena where a recognizable delayed effect is due to a specific cause—impulse on the system. Dielectric relaxation phenomena, the time taken for dipoles to orient themselves in an electric field, are another example. Electronic and nuclear relaxation phenomena are also widely used to study properties of molecules. The processes discussed here have been specifically developed by Eigen and de Maeyer (1963) under the name 'chemical relaxations' because the delay in response to perturbation is due to the finite time taken by one or a series of elementary chemical processes, for example

$$E + S \rightleftharpoons X_1 \rightleftharpoons X_2 \ldots \rightleftharpoons ES$$

A single exponential decay in a relaxation phenomenon does not prove that there is only a single reaction step, but it is consistent with there only being the two states of the reactants populated to a significant extent. Degeneracy of the relaxation times and insufficient experimental accuracy can result in difficulties of obtaining more than one relaxation time from multi-step processes. In theory there are $n - 1$ relaxation times for n identifiable states of the reactants. In the exceptional circumstances when each of a number

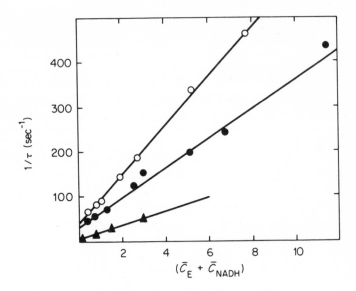

Fig 52   Plots of the reciprocal relaxation times against concentrations for the system lactate dehydrogenase + NADH in 0·1 M sodium phosphate, with ionic strength of 0·3 M with NaCl. *Lines* were drawn by the method of least squares. ●, 20°, pH 6·00; ○, 20°, pH 8·00; ▲, 3°, pH 6·00 (from Heck, 1969)

of relaxation times differs from all the others by more than an order of magnitude, the relaxations are said to be uncoupled from each other. In a system with just two relaxation times uncoupling is not uncommon and leads to great simplification both in the evaluation of the relaxation times and in the complexity of the expressions for the relaxation times in terms of individual rate constants. This will be apparent from some specific examples.

The algebraic procedure for deriving the relaxation times is similar to that used for steady-state kinetics. A set of differential equations is written, each giving the time dependence of the deviation from equilibrium of one of the intermediates. With the help of matrix algebra the relaxation times are derived as the roots of the secular equation. While the simple one-step reaction is described by

$$\Delta C = \Delta C° \, e^{-t/\tau}$$

where $\Delta C°$ is the concentration deviation at $t = 0$, for a series of coupled reactions a system of first-order differential equations has to be solved

$$-\frac{d\Delta C_i}{dt} = \sum_{j=1}^{n} a_{ij}\Delta C_j$$

(a_{ij}s are functions of the rate constants and equilibrium concentrations). The system of equations can be written as

$$-\frac{d\{C\}}{dt} = \{A\}\{C\}$$

where $\{C\}$ is the column matrix $(\Delta C_1, \Delta C_2, \ldots \Delta C_n$, the concentration deviation from equilibrium of the reaction intermediates) and $\{A\}$ is the n by n matrix of the coefficients a_{ij}. The solution of the above equation is

$$\{C(t)\} = e^{-\{A\}t}\{C(0)\} = [\sum e^{-\lambda_j t} f_j(\{A\})]\{C(0)\}$$

where the f_js are characteristic matrices of A and the λ_js are the eigenvalues obtained by solving the determinant

$$\begin{vmatrix} a_{11} - \lambda & a_{12} & \ldots & a_1n \\ a_{21} & a_{22} - \lambda & \ldots & a_2n \\ \vdots & \vdots & \vdots & \vdots \\ a_{nl} & \ldots & \ldots & a_{nn} - \lambda \end{vmatrix} = 0$$

In a linear relaxation process the integrated rate equation is the sum of exponentials or terms of first-order rate equations. In the present case $1/\lambda_j = \tau_j$ and a spectrum of relaxation times is obtained. In principle each relaxation time is a function of the equilibrium concentrations of all the intermediates. It will be seen, however, that in practice steps can be uncoupled from the rest of the system. The discussion in the above paragraph is close to that of Amdur and Hammes (1966).

The response to a temperature perturbation of a system in which a number of coupled reactions are at equilibrium proceeds with a set of coupled relaxation times. In the interaction between enzymes and substrates the two-step mechanism

$$\text{E} + \text{S} \underset{}{\overset{\kappa_1}{\rightleftharpoons}} \text{ES} \underset{}{\overset{\kappa_2}{\rightleftharpoons}} \text{E*S}$$

is of particular interest. The problem to be solved is the solution of the rate equations in terms of relaxation times. The rate equations for this model are

$$\frac{dC_E}{dt} = -k_1 C_E C_S + k_{-1} C_{ES}$$

$$\frac{dC_{E*S}}{dt} = k_2 C_{ES} - k_{-2} C_{E*S}$$

The choice of the intermediates which are used for the rate equations depends on which of them are observed by the monitoring technique. The conditions for linearization and conservation of matter are used as for the first example to obtain

$$\frac{d\Delta C_E}{dt} = -[k_{-1} + k_1(\bar{C}_E + \bar{C}_S)]\Delta C_E - k_{-1}\Delta C_{E*S}$$

$$\frac{d\Delta C_{E*S}}{dt} = -k_2 \Delta C_E - (k_2 + k_{-2})\Delta C_{E*S}$$

From this we can write

$$\begin{vmatrix} [k_{-1} + k_1(C_E + C_S)] - \dfrac{1}{\tau} & k_{-1} \\[2mm] k_2 & (k_{-2} + k_2) - \dfrac{1}{\tau} \end{vmatrix} = 0$$

This can be written as

$$\left(\frac{1}{\tau}\right)^2 - [k_{-1} + k_1(\bar{C}_E + \bar{C}_S) + k_2 + k_{-2}]\frac{1}{\tau}$$

$$+ [k_{-1}k_{-2} + k_1(k_2 + k_{-2})(C_E + C_S)] = 0$$

The two solutions of this quadratic equation give the reciprocals of the two relaxation times τ_1 and τ_2.

$$\frac{1}{\tau_1} + \frac{1}{\tau_2} = k_{-1} + k_1(\bar{C}_E + \bar{C}_S) + k_2 + k_{-2} \tag{8.15}$$

$$\frac{1}{\tau_1}\frac{1}{\tau_2} = k_{-1}k_{-2} + k_1(k_2 + k_{-2})(\bar{C}_E + \bar{C}_S) \tag{8.16}$$

If the two steps consist of a very fast binding step and a much slower isomerization step, then

$$\frac{1}{\tau_1} \gg \frac{1}{\tau_2}$$

and the two relaxation times can be considered to be uncoupled. In such cases the rapid binding step can be treated as independent and

$$\frac{1}{\tau_1} = k_{-1} + k_1(\bar{C}_E + \bar{C}_S) \tag{8.17}$$

The isomerization step is coupled to the concentration dependence of the previous step

$$\frac{1}{\tau_2} = \frac{k_{-1}k_{-2} + k_1(k_2 + k_{-2})(\bar{C}_E + \bar{C}_S)}{k_{-1} + k_1(C_E + C_S)}$$

$$= k_{-2} + \frac{k_2}{1 + K_1/(\bar{C}_E + \bar{C}_S)} \tag{8.18}$$

where $K_1 = k_{-1}/k_1$. It follows that

$\dfrac{1}{\tau_2}$ approaches k_{-2} when $C_E + C_S \ll K_1$

$\dfrac{1}{\tau_2}$ approaches $k_{-2} + k_2$ when $C_E + C_S \gg K_1$

On the other hand, the fast relaxation characterized by $1/\tau_1$ will increase linearly with $(C_E + C_S)$.

The analysis of such a two-step mechanism by stopped-flow measurements was discussed in Section 8.2.3. The rate of appearance of E*S after mixing enzyme and substrate can only be interpreted directly if $k_1(C_E + C_S) \gg (k_2 + k_{-2})$ and $C_S \gg C_E$. The observed first-order rate constant k_{obs} becomes equal to $k_2 + k_{-2}$, when $C_S \gg K_1$. The equation can also be linearized by the condition $C_E \gg C_S$.

It is of great interest to distinguish between the above mechanism of a substrate-induced conformation change and the pre-equilibration between two forms of the enzyme, with only E* binding substrate:

$$E \underset{k_{-1}}{\overset{k_1}{\rightleftharpoons}} E^* \underset{\pm S}{\overset{k_2}{\rightleftharpoons}} E^*S$$

The derivation of the relaxation times of such a system is taken from Halford (1970). The rate equations for this model are

$$\frac{dC_E}{dt} = k_{-1}C_{E^*} - k_1C_E$$

$$\frac{dC_{E^*S}}{dt} = k_2C_{E^*}C_S - k_{-2}C_{E^*S}$$

The introduction of conditions for linearization and conservation of mass results in

$$\frac{d\Delta C_E}{dt} = -(k_1 + k_{-1})\Delta C_E - k_{-1}\Delta C_{E*S}$$

$$\frac{d\Delta C_{E*S}}{dt} = k_2 C_S \Delta C_E - (k_{-2} + k_2)\{C_{E*} + C_S\}C_{\Delta E*S}$$

Hence we can write

$$\begin{vmatrix} (k_1 + k_{-1}) - \dfrac{1}{\tau} & k_{-1} \\[2ex] k_2 C_S & (k_{-2} + k_2)\{C_{E*} + C_S\} - \dfrac{1}{\tau} \end{vmatrix} = 0$$

and

$$\left(\frac{1}{\tau}\right)^2 - (k_1 + k_{-1} + k_{-2} + k_2\{C_{E*} + C_S\})\frac{1}{\tau}$$

$$+ k_2(k_1\{C_{E*} + C_S\} + k_{-1}C_{E*}) + k_{-2}(k_1 + k_{-1}) = 0 \quad (8.19)$$

The solutions $1/\tau_1$ and $1/\tau_2$ to this quadratic equation can be written as

$$\left(\frac{1}{\tau}\right)^2 - \frac{1}{\tau_1} + \frac{1}{\tau_2}\frac{1}{\tau} + \frac{1}{\tau_1}\frac{1}{\tau_2} = 0 \quad (8.20)$$

Comparison of equations (8.19) and (8.20) shows that

$$\frac{1}{\tau_1} + \frac{1}{\tau_2} = k_1 + k_{-1} + k_{-2} + k_2\{C_{E*} + C_S\}$$

$$\frac{1}{\tau_1}\frac{1}{\tau_2} = k_2(k_1\{C_{E*} + C_S\} + k_{-1}C_{E*}) + k_{-2}(k_1 + k_{-1})$$

If the relaxations are sufficiently well separated we can again use the procedure for deriving expressions for uncoupled processes. If the substrate binding process is much faster than the isomerization $E \rightleftharpoons E^*$ (i.e. $1/\tau_2 \gg 1/\tau_1$) it follows that

$$\frac{1}{\tau_2} = k_{-2} + k_2(C_{E*} + C_S) \quad (8.21)$$

and

$$\frac{1}{\tau^1} = \frac{\dfrac{1}{\tau_1}\dfrac{1}{\tau_2}}{\dfrac{1}{\tau_2}} = \frac{k_2(k_1\{C_{E*} + C_S\} + k_{-1}C_{E*}) + k_{-2}(k_1 + k_{-1})}{k_{-2} + k_2(C_{E*} + C_S)} \quad (8.22)$$

$$\frac{1}{\tau_1} = k_1 + \frac{k_{-1}}{1 + C_S/(C_{E^*} + K_2)} \tag{8.23}$$

where $K_2 = k_{-2}/k_2$

$\dfrac{1}{\tau_1}$ tends towards $k_1 + k_{-1}$ when $C_S \ll (C_{E^*} + K_2)$

$\dfrac{1}{\tau_1}$ tends towards k_1 when $C_S \gg (C_{E^*} + K_2)$

The two mechanisms of (substrate-induced isomerization before or after substrate binding) can be distinguished by their different response to changes in substrate concentration. Comparison of the expressions for the relaxation times of the first-order isomerization steps of the two mechanisms (τ_2 in the case of the substrate-induced isomerization and τ_1 in the case of isomerization before substrate binding) shows that in the first case as C_S is increased $1/\tau_2$ increases to a maximum value, in the second case increasing C_S results in a decrease in $1/\tau_1$. Furthermore, the expression for $1/\tau_2$ in the first mechanism contains a term $(C_E + C_S)$ and when K_1 is a very small concentration independent relaxation times are observed. In the second mechanism, however, the expression for $1/\tau_1$ contains C_E and C_S as a fraction and, while it may be impossible in practice to vary $(C_E + C_S)$ with respect to K_1, it should always be possible to vary C_S with respect to $(C_{E^*} + K_1)$.

The above argument shows that the relaxation technique can distinguish between the two alternative mechanisms of two-step enzyme substrate complex formation. The distinction through C_S dependence of the reciprocal relaxation time becomes difficult when $k_1 \gg k_{-1}$ (of the second mechanism). In that case, however, it is possible to distinguish the two mechanisms through transient kinetic experiments.

When one considers a mechanism one should always write down a general scheme, as for instance:

$$S + E \rightleftharpoons E^* + S$$
$$\updownarrow \quad\quad \updownarrow$$
$$ES \rightleftharpoons E^*S \rightarrow \text{Reaction}$$

In theory it is obvious that in either of the two-step mechanisms considered above all four forms of the enzyme can exist. From a thermodynamic point of view it is of interest to obtain information about all four equilibrium constants even if from a kinetic point of view the reaction proceeds entirely clockwise or entirely anticlockwise. The difference between the thermodynamic and kinetic point of view arises from the fact that an equilibrium constant of 100 or 0·01 can be measured while an alternative reaction path of 1 part in 100 will rarely be detected. This problem arises also in the more complex equilibria discussed in Section 8.3.4.

The principles outlined above should enable one to derive expressions for relaxation times for any other mechanism of enzyme–substrate interaction. Hammes and his colleagues (del Rosario and Hammes, 1970, and Hammes and Hurst, 1970) have derived equations for the relaxation of systems involving the interaction of an enzyme with two substrates and of an enzyme with a substrate and metal ion. For a very clear elementary treatment of relaxation phenomena Hague (1971) should be consulted.

8.3.2. The Evaluation of Relaxation Times and Rate Constants

The experimental record obtained from the observation of the relaxation of a single-step perturbation can be represented by the sum of exponentials. The first problem is to evaluate the individual exponentials from the record of an experiment under a particular set of conditions. If a system has a single relaxation, $1/\tau$ is simply obtained as the slope of $\ln C_X$ against t, where C_X is the concentration of the observed intermediate. If there are two or more relaxations it depends on their separation which of three methods can be used to evaluate the individual relaxation times:

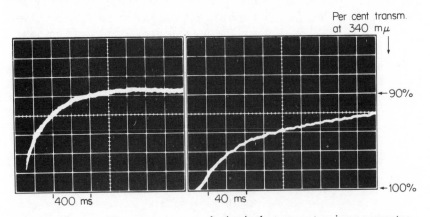

Figure 53   A rapid temperature perturbation in the temperature jump apparatus of the equilibrium system:

$$\text{Ethanol} + \text{NAD} \rightleftharpoons \text{Acetaldehyde} + \text{NADH} + \text{H}^+$$

in the presence of 40 M (in enzyme sites) horse liver alcohol dehydrogenase. The reaction was carried out at pH 6·5 in 0·067 M phosphate buffer and ethanol + acetaldehyde as well as NAD + NADH were 10 m M

In the fast scan the slower relation, which has a relaxation time of 200 ms, appears as a straight line after the fast relaxation, which has a relaxation time of
20 ms

(1) If the relaxation times differ by more than an order of magnitude the individual exponentials can be evaluated from records obtained at different scanning speeds (see Figure 53).

(2) If two relaxations differ by a factor of 3 or more, they can be evaluated in a manner similar to that used for parallel first-order reaction (see Figure 24). The slower exponential is evaluated from a record of the last part of the reaction and subtracted from the whole record. The faster exponential can then be calculated from the residue.

(3) If either the number or closeness of the relaxation times makes the above methods inapplicable, a computer program has to be used to fit the exponentials to the experimental data.

The simplest reaction, a simple unimolecular isomerization, results in a relaxation with a single concentration-independent constant ($1/\tau = k_1 + k_{-1}$) and requires no further consideration. The next model to be discussed is the single-step, second-order association, first-order dissociation with a single concentration-dependent relaxation $[1/\tau = (C_S + C_E)k_1 + k_{-1}]$. In such a system the evaluation of the rate constants from relaxation experiments is achieved by plotting the sum of the concentrations of free enzyme sites and free substrate against $1/\tau$. If the assumed mechanism is correct, a straight line will be obtained with slope k_1 and ordinate intercept k_{-1}. This procedure is illustrated in Figure 52. If the equilibrium constant

$$K = \frac{\bar{C}_E \bar{C}_S}{\bar{C}_{ES}} = \frac{k_{-1}}{k_1}$$

is not known the following reiterative procedure has to be used. An equilibrium constant is assumed for the calculation of free enzyme and substrate concentration from the total concentration. From the first plot a value for k_1 and k_{-1} is obtained and an equilibrium constant is calculated from these rate constants. This equilibrium constant is used to calculate free reactant concentrations for the next plot and so on. The equilibrium constant can also be evaluated from the dependence of the relaxation amplitude on the reactant concentrations.

The interpretation of plots of relaxation times against reactant concentrations for more complex mechanisms was discussed with the derivations presented in the previous section. The number of relaxations observed are equal to or smaller than those actually possible. It depends on the equilibrium constants and the range of concentrations over which reactions can be followed whether a reaction can be studied under conditions which permit the resolution of all possible relaxations. The method of observation also determines the number of observed relaxations. Usually the optical observation of the progress of the reaction provides a record of the changes in concentration of one reactant only. The number of relaxations which can be seen depends not only on their relative magnitude but also on the position of the observed reactant within the scheme. For instance, in a mechanism

$$A + B \underset{\longleftarrow}{\overset{k_1}{\rightleftharpoons}} C \underset{\longleftarrow}{\overset{k_2}{\rightleftharpoons}} D$$

if $1/\tau_1 \gg 1/\tau_2$ and D is the observed intermediate only one relaxation with time constant $1/\tau_2$ will be seen. However, the time constant $1/\tau_2$ will depend on the concentrations of A and B and when C_A and C_B are small $1/\tau_1$ will become of comparable magnitude to $1/\tau_2$ and both relaxations will be observed. If C is the observed intermediate both relaxations should be seen at all concentrations C_A and C_B.

8.3.3. The Perturbation of Equilibria of Enzyme Reactions

In any system the condition for being able to study the perturbation of equilibria is that at equilibrium all reactants occur at significant concentrations. When the equilibrium constant of a reaction predicts that it will go essentially to completion, some partial reactions can be used to study individual steps but the catalytic reaction cannot be investigated by equilibrium-relaxation techniques. In such cases it is possible, in principle, to study relaxations of the perturbed steady state in a combined flow–temperature apparatus. This technique has so far had very limited application. Some of the potentialities and limitations of this method are discussed by Erman and Hammes (1966) and Gutfreund (1971).

For a sequential pathway the minimum number of intermediates, that is the simplest meaningful mechanism, is represented by

$$E + S \overset{k_1}{\rightleftharpoons} ES \overset{k_2}{\rightleftharpoons} EP \overset{k_3}{\rightleftharpoons} E + P$$

In the case of substituted enzyme mechanisms the equilibrium of partial reaction of the type

$$E + AB \rightleftharpoons EAB \rightleftharpoons EA + B$$

can be studied. Extensive studies of relaxations of equilibria of transaminase reactions have been reviewed by Fasella (1967).

The three relaxation times of the sequential mechanism are the roots of the third-order equation obtained from the three linearized rate equations. The cubic equation can only be solved by approximation. For one form of approximate treatment and references to others, the article by Hammes and Schimmel (1970) should be consulted. It is both unnecessarily difficult and dangerous to interpret relaxation phenomena of overall enzyme-catalysed reactions without the use of information available from steady-state and transient kinetics.

It may be instructive to use the reaction of heart lactate dehydrogenase to illustrate how one might interpret its chemical relaxations. From detailed studies of the steady state and transient kinetics of this enzyme the minimum number of steps of the reaction are represented by

$$E \underset{}{\overset{k_1}{\rightleftharpoons}} ENAD \underset{}{\overset{k_2}{\rightleftharpoons}} E^{NAD}_{Lact} \underset{}{\overset{k_3}{\rightleftharpoons}} E^{*NAD}_{Lact} \underset{}{\overset{k_4}{\rightleftharpoons}} E^{*NADH}_{Pyr} \underset{}{\overset{k_5}{\rightleftharpoons}} E^{NADH}_{Pyr}$$
$$\underset{}{\overset{k_6}{\rightleftharpoons}} ENADH \underset{}{\overset{k_7}{\rightleftharpoons}} E$$

Figure 54 shows the relaxation of this reaction under one particular condition. Two distinct relaxation times are observed and more may be found when the system is examined under a wider range of conditions. The fast relaxation could be due to the hydride transfer (Step 4) or to some step

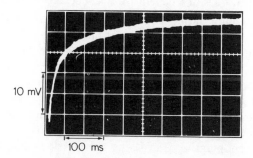

Figure 54 The chemical relaxation of the equilibrium of the lactate dehydrogenase reaction can be resolved into two relaxation times. The temperature perturbation from 0 to 4·5° was obtained by a 30 kV discharge. The concentration of pig heart enzyme was 2 mg/ml and the initial concentration of added substrate was NAD^+, 10 mM, lactate, 2 mM, in 0·1 M phosphate buffer pH 7·8

closely coupled to it. A decision on this can be obtained from carrying out the experiment with deuterium-labelled substrate and by comparing the rates with both isotopes with the hydride transfer rate determined by the transient technique. The transient method has been illustrated for the analogous mechanism of liver alcohol dehydrogenase and some information on isotope effects on transients of lactate dehydrogenase reactions has been obtained (see page 175). It is known from stopped-flow experiments that the rate-limiting step in the reaction from lactate and NAD to pyruvate and NADH is a conformation change which controls the release of NADH from its enzyme complex. It is not clear whether this slow step occurs before or after the release of pyruvate. The slow relaxation is likely to be associated with the rate of release of free NADH and the effect of pyruvate concentration on the slow relaxation time can fill in this particular part of the jigsaw puzzle.

The purpose of this discussion is to emphasize that no one method is omnipotent at solving reaction mechanisms, but the combination of all the kinetic methods and tricks of changes in substrate reactivity and isotopic substitution should be employed to describe the complete reaction path. It is just as important to use a wide range of observation techniques to observe different intermediates as it is to use a wide range of forcing methods to initiate or perturb the reactions at different points.

8.3.4. Kinetic Observations of Cooperative Phenomena

In Section 8.3.1 conformation changes in response to substrate binding were discussed for reactions at independent binding sites. In Sections 4.3.1 to 4.3.6 the binding equilibria of interacting (cooperative) binding sites were described together with a number of possible mechanisms which could be responsible for cooperative binding phenomena. It now remains to be shown how kinetic investigations might distinguish between different mechanisms. As will be seen, some of the structural changes are quite slow and can be investigated more easily with stopped-flow methods than with relaxation techniques. Some cooperative transitions are so slow that they can be observed in a conventional spectrophotometer. In the present section kinetic information on cooperative phenomena will be discussed irrespective of the experimental or theoretical method of analysis.

We shall first discuss the formal problem to be solved and the use one might make of this formal description. The detailed analyses will be discussed by describing the properties of some systems which have been investigated. As pointed out in Section 4.3.6, the proposals by Monod and his colleagues stimulated many investigators to try and design experiments which could provide evidence for or against the model in which all sites (or subunits) are in one of two forms with ligand-binding affinities K_T and K_R, respectively. Koshland and his colleagues proposed a model with a series of different binding constants in a sequential process of binding and conformation changes. Equilibrium binding studies can provide useful practical information and they can provide evidence for a sequential process when there is negative cooperativity (see p. 156). Equilibrium studies cannot prove one or other mechanism of positive cooperativity. The particular power of kinetic methods in this field is their ability to prove the existence of a *minimum* number of steps. The Monod all-or-none mechanism is very restrictive, if too many distinct steps are observed it can be eliminated. If the number of steps observed is just the right number, this does not prove the all-or-none mechanism but it suggests that it might apply.

The difference between the Monod model and the Koshland model has been illustrated schematically (p. 91) but we are now able to discuss its significance with reference to some of the kinetic concepts treated in earlier sections of this chapter. Just as we were interested in using kinetics to analyse the sequence of events in the series of steps involved in enzyme catalysis, we now wish to see how far kinetics can help to determine the series of events during regulation. We are only going to concern ourselves with the 'relatively' simple case of a single ligand binding ‘under conditions of positive cooperativity. For the Koshland model of such a case one should use the scheme shown in Figure 55.

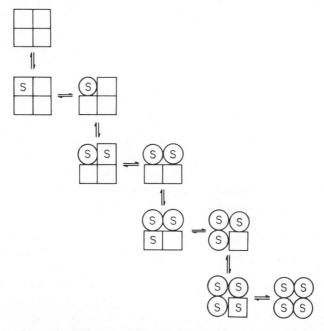

Figure 55   The induced-fit model of Koshland, Nemethy and Filmer (1966) in which the conformation changes of each subunit occur progressively after ligand binding

It is proposed that there is a substrate-induced conformation change, which only occurs after the substrate is bound to the T-form (low affinity form) of the subunit. The initial binding to the four T-units has equivalent intrinsic constants but the subsequent conformation changes (resulting in the overall binding constant to the R-form) depend on interaction, i.e. the number of $T \rightarrow R$ transitions which have already occurred within one polymer. If such a scheme were strictly followed one second-order process and four first-order processes should be observed in a detailed kinetic investigation. Much more complex relaxation spectra can be observed in a more general mechanism.

Equilibrium studies on their own cannot even result in the distinction between the two sequences of substrate-induced conformation changes in non-cooperative systems, which have been analysed kinetically in Section 8.3.1. Halford (1970) has shown that a small change in condition of solution of alkaline phosphatase (*E. coli*) can change the sequence of events from a substrate-induced conformation change to a pre-equilibrium of two conformations, only one of which will bind the substrate. It has not yet been shown whether the above Koshland scheme can be distinguished from one in which one goes one step to the right first and then one step down and so on.

Monod et al. proposed that there are only two conformations of the polymer with all units in the T-state or all units in the R-state. There is an

equilibrium between the unliganded, as well as between the mono, di, tri, etc., liganded T-form and R-form. In such a system the substrate–polymer interaction will be described by two second-order processes (binding to the T- and R-form, respectively) and one first-order transition. This simple model, which can be described by three equilibrium constants or the rate constants of three reversible processes was not proposed simply as the mathematical limit of an efficient cooperative process. The model was suggested as a logical design of polymeric proteins as efficient switching systems.

The real difficulty in finding out whether the all-or-none model or the sequential model operates, in any one particular system exhibiting positive cooperativity, is in the accuracy required from the experimental methods. Two separate rate constants which only differ by a factor of two are difficult to distinguish and intermediate states which are at very low concentrations are difficult to detect but may be on the direct pathway.

One aim of kinetic investigations is to define as many as possible of the intermediates and then to characterize the structural differences between these intermediates by spectroscopic, potentiometric and other continuous monitoring techniques. Protein molecules exhibit a fair amount of flexibility and when relaxation and other kinetic experiments demonstrate conformation changes, the different states do not necessarily correspond to forms of different affinities or catalytic activities. It is always necessary to correlate rates of conformation changes with rates of changes of some specific function of the enzyme. The subsequent discussion is restricted to cooperative systems with intrinsically identical binding sites. The most intensively investigated cooperative protein, haemoglobin, probably does not fall into this class. The α and β subunits certainly have different structures and undergo different conformation changes during ligand binding (see discussion and references quoted in Section 4.3.3). Haemoglobin is a very nice system to study because it can be obtained so easily in large quantities, and can be investigated by so many different methods. However, the kinetic complexity of the reactions of this multifunctional carrier have so far not been elucidated unambiguously.

The first and most detailed analysis of the kinetics of cooperative substrate binding was carried out by Kirschner et al. (1966) on the system yeast glyceraldehyde 3-phosphate dehydrogenase (GAPDH) and NAD^+. At alkaline pH this enzyme undergoes a temperature-dependent structure change which is linked to NAD^+ binding. This reaction is probably not of biological significance, but it serves as an example for a process which is clearly compatible with the Monod model. The enzyme has four NAD binding sites and the spectral change (the Racker band near 365 mμ) during the complex formation can be used for kinetic investigations.

This system, like most other ligand binding processes, can be treated in terms of rapid binding followed by a slow conformation change. In such a case two relaxation times are observed for association with sites in the R-form and T-form, respectively. These two concentration dependent processes with time constants τ_R and τ_T are uncoupled if one of two conditions holds: first when one relaxation time is much shorter than the other and secondly when the ligand concentration is very large compared with the binding site concentration. If the substrate concentration is large (i.e. buffered) $1/\tau_R$ and $1/\tau_T$ will increase linearly with substrate concentration. In the case of the GAPDH–NAD^+ reaction at 40° and pH 8·5 two fast relaxations with $1/\tau$ dependent on C_{NAD} are observed (Figure 56(a) and (b)). The two relaxation times and the two pairs of rate constants were evaluated as shown in Section 8.3.2 and it was found that for association with the T-form

$$k_1 = 1·37 \times 10^6 \text{M}^{-1} \text{sec}^{-1}$$
$$k_{-1} = 0·21 \times 10^3 \text{sec}^{-1}$$
$$k_{-1}/k_1 = 1·5 \times 10^{-4} \text{ M}$$

and for association with the R-form

$$k_1 = 1·9 \times 10^7 \text{M}^{-1} \text{sec}^{-1}$$
$$k_{-1} = 1·0 \times 10^3 \text{sec}^{-1}$$
$$k_{-1}/k_1 = 5·3 \times 10^{-5} \text{ M}$$

Figure 56(c) shows that this system has a third much longer and easily isolated relaxation time. The decision about the mechanism was made on the characteristic concentration dependence of this third relaxation time. The reciprocal relaxation time $1/\tau_C$ is attributed to the conformation change, it is independent of enzyme concentration and decreases with the nth power of $1/(1 + C_S/K_R)$, where n is the cooperativity. The rate constants k_1 (for $T \rightarrow R) = 0.18 \, sec^{-1}$ and k_{-1} (for $R \rightarrow T) = 5.5 \, sec$ describe the equilibrium of the unliganded protein $L_0 = R_0/T_0 = k_{-1}/k_1 = 30$. A detailed

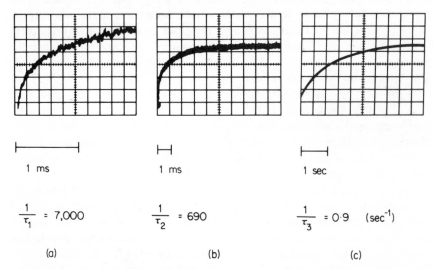

$$\frac{1}{\tau_1} = 7{,}000 \qquad \frac{1}{\tau_2} = 690 \qquad \frac{1}{\tau_3} = 0.9 \quad (sec^{-1})$$

(a)   (b)   (c)

Figure 56   The relaxations of the system yeast glyceraldehyde 3-phosphate dehydrogenase-NAD after temperature perturbations in solutions containing: enzyme, 20 mg/ml (1.4×10^{-4} M in tetramer); NAD^+, 6×10^{-4} M; pyrophosphate pH 8.5, 0.05 M; EDTA, 5×10^{-3} M. The temperature perturbations were from 40° to approximately 45°. The records taken at three different sweep speeds show (a) and (b) the association equilibria of enzyme with NAD^+ and (c) the transformation of the T-form (lower affinity for NAD^+) to the R-form (higher affinity for NAD^+) (Kirschner, Eigen, Bittman and Voight, 1966)

treatment of the algebra describing the above model and its relation to the more general sequential model is given by Eigen (1967).

In most cases cooperative transitions are treated in terms of increases or decreases in ligand affinity during the conformation change. Another interesting aspect of cooperative conformation changes is the transition between catalytically inactive to active enzyme. In the case of yeast glyceraldehyde phosphate dehydrogenase, for instance, it was found that the mixing of enzyme with NAD^+ and the other substrates results in a slow activation process. The time course of this activation process is similar to that of the $T \rightarrow R$ transition. If $ENAD^+$ is mixed with the other substrates full activity is observed from the first moment of observation. In the case of yeast pyruvate kinase and rabbit muscle phosphofructokinase one can also observe slow transitions corresponding to the rates of activation or inhibition on addition of substrate or inhibitor. The rate constants of activation processes can be evaluated simply graphically:

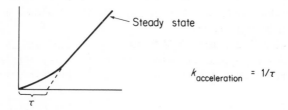

In recent investigations of cooperative transitions the catalytic process was often somewhat neglected. There is so far not nearly enough detail available to correlate the transitions with the different functions of the enzymes and with particular changes in protein structure.

Aspartokinase 1-homoserine dehydrogenase undergoes $T \rightarrow R$ transformations linked to binding of aspartate and K^+ ions. The complexities of the subunit structure and dual activity of this enzyme were discussed in Section 4.3.5. In the relaxation studies of Janin and Iwatsubo (1969) the structural transitions of the enzyme were followed directly through differences in the absorption and fluorescence spectra of different conformations of the protein. They concentrated on the study of these transformations rather than on the ligand binding process itself. The studies of Janin and Iwatsubo serve at present mainly as an example of the variety of ways in which a cooperative process can be investigated. The mechanism of the complex problem has not yet been elucidated.

With aspartokinase 1-homoserine dehydrogenase some of the relaxations are sufficiently slow to be conveniently investigated in a stopped-flow spectrophotometer. At moderately low aspartate concentration this ligand will only bind significantly to the R-form. Mixing enzyme with aspartate permits the evaluation of the rate of the $T_0 \rightarrow R_0$ transition. The rate constant of 15 sec^{-1} for this process and 150 sec^{-1} for the reverse reaction were calculated from these experiments and the equilibrium $L = T_0/R_0 = 10$, determined spectrophotometrically. A special point must be remembered in the interpretation of these experiments as well as of those on glyceraldehyde phosphate dehydrogenase. The conformation change is uncoupled inasmuch as it is rate limiting for the kinetic process observed. However, the reversibility of the conformation change is affected by ligand binding to the R-form and the much slower $R \rightarrow T$ transformation of the liganded enzyme. Algebraically this is expressed in the equations derived by Eigen (1967).

However, as would be expected for such a complex protein, temperature jump relaxations indicate additional conformational transition. The transition T → R has $\Delta H < 0$, which means that an increase in temperature shifts the equilibrium towards an increase in the T-form. By this technique an additional faster conformational transition was detected in studies of homoserine dehydrogenase.

The examples of relaxation studies on enzymes described in this and the previous section are typical of many in progress at this time. A more comprehensive list of recent investigations will be found in a review (Gutfreund, 1971).

Another application of relaxation techniques to biological problems, which is of topical interest, is the study of cooperative structure changes (with or without ligand binding) of nucleic acids and other hydrogen bonded polymers. References to various investigations in this field are found in the papers by Eigen (1967) and Schwarz (1969). The types of problems which are being actively investigated are basic to the elucidation of the relation between structure, information contents (capability) and evolution of biological macromolecules. Helical and other conformations of polymers are highly cooperative because the loops of the structure are always stabilized by a number of interactions. Once one of these contacts (often hydrogen bonds) is formed, the probability for the formation of the others is considerably enhanced. Similarly, binding of small molecules to nucleic acids is highly cooperative and it is fitting to end a discussion on the contribution of relaxation techniques to biology with a further comment on Eigen's work in this field. Kinetic studies in the Göttingen laboratory have shown that the triplet code optimizes the relation between information, which increases with the size of codon, and the rate of dissociation of the codon–anticodon complex, which decreases with the size of the codon. The conclusion from rate measurements was that larger codons than triplets would dissociate too slowly for the efficient reading and replication process (see Eigen, 1967).

REFERENCES

Adair, G. S. (1925). *J. biol. Chem.*, **63**, 529.

Alberty, R. A. (1953). *J. Am. Chem. Soc.*, **75**, 1925 and 1928.

Alberty, R. A. (1969). *J. Am. Chem. Soc.*, **91**, 3899 and *J. biol. Chem.*, **244**, 3290.

Amdur, I. and Hammes, G. G. (1966). *Chemical Kinetics*, McGraw-Hill, New York, Chap. 6.

Antonini, E. and Brunori, M. (1970). *Ann. Rev. Biochem.*, **39**, 977.

Antonini, E., Schuster, T. M., Brunori, M., and Wyman, J. (1965). *J. biol. Chem.*, **240**, PC 2262.

Antonini, E., Wyman, J., Moretti, R., and Rossi-Fanelli, A. (1963). *Biochim. Biophys. Acta*, **71**, 124.

Barnard, E. A. (1969). *Ann. Rev. Biochem.*, **38**, 677.

Bates, R. G. and Schwarzenbach, G. (1955). *Helv. Chim. Acta*, **38**, 699.

Bell, R. M. and Koshland, D. E. (1971). *Science*, **172**, 1253.

Bender, M. L. (1971). *Mechanisms of Homogeneous Catalysis from Protons to Proteins*, Wiley, New York.

Bender, M. L. and Kezdy, F. J. (1965). *Ann. Rev. Biochem.*, **34**, 49.

Bender, M. L., Pollock, E. J., and Neveu, M. C. (1962). *J. Am. Chem. Soc.*, **84**, 595.

Bernhard, S. A. (1956). *J. biol. Chem.*, **218**, 961.

Blake, C. C. F., Koenig, D. F., Mair, G. A., North, A. C. T., and Phillips, D. C. (1965). *Nature*, **206**, 757.

Bohr, C., Hasselbalch, K. A., and Krogh, A. (1904). *Skand. Arch. Physiol.*, **16**, 402.

Briggs, G. E. and Haldane, J. B. S. (1925). *Biochem. J.*, **19**, 383.

Bright, H. J. and Gibson, Q. H. (1967). *J. biol. Chem.*, **242**, 494.

Bruice, T. C. and Benkovic, S. J. (1966). *Bioorganic Mechanisms*, Benjamin, New York.

Chance, B. (1943). *J. Biol. Chem.*, **151**, 553.

Chance, B. (1963). *Rates and Mechanisms of Reactions*, Wiley, New York, p. 1314.

Chance, B. (1967). *Methods in Enzymology*, **10**, 641.

Chance, B. (1967). *Nobel Symposium*, **5**, 437.

Chipman, D. M. and Schimmel, P. R. (1968). *J. biol. Chem.*, **243**, 3771.

Cohn, M. (1970). *Quarterly Rev. Biophys.*, **3**, 61.

Dalziel, K. (1957). *Acta Chem. Scand.*, **11**, 1706.

Dalziel, K. (1969). *Biochem. J.*, **114**, 547.

Dalziel, K. and Dickinson, F. M. (1966). *Biochem. J.*, **100**, 34 and 491.

Darrow, R. A. and Colowick, S. P. (1962). *Methods in Enzymology*, **5**, 226.

Darvey, I. G. (1968). *J. Theoret. Biol.*, **19**, 215.

del Rosario, E. J. and Hammes, G. G. (1970). *J. Amer. Chem. Soc.*, **92**, 1750.

Dickerson, R. E. and Geis, I. (1969). *The Structure and Action of Proteins*, Harper and Row, New York.

Diebler, H., Eigen, M., Ilgenfritz, G., Maas, G., and Winkler, R. (1969). *Pure and Applied Chem.*, **20**, 93.

Dixon, M. and Webb, E. C. (1964). *Enzymes*, 2nd Edn, Longmans, Green, London.

229

Drenth, J., Jansonius, J. N., Koekoek, R., Sluyterman, L. A. A., and Wolthers, B. G. (1970). *Phil. Trans. Roy. Soc. Lond.*, **B257**, 231.

Edsall, J. T. and Wyman, J. (1958). *Biophysical Chemistry*, Academic Press, New York.

Eigen, M. (1963). *Angew. Chemie*, **75**, 489.

Eigen, M. (1967). *Nobel Symposium*, **5**, 333.

Eigen, M. (1968). *Quarterly Rev. Biophys.*, **1**, 3.

Eigen, M. and de Maeyer, L. C. (1963). *Rates and Mechanisms of Reactions*, Wiley, New York, p. 895.

Engel, P. C. and Dalziel, K. (1969). *Biochem. J.*, **115**, 621.

Erman, J. E. and Hammes, G. G. (1966). *Rev. Sci. Instrum.*, **37**, 746.

Fasella, P. (1967). *Ann. Rev. Biochem.*, **36**, 185.

Findlay, D., Mathias, A. P., and Rabin, B. R. (1962). *Biochem. J.*, **85**, 139.

Finlayson, B., Lymn, R. W., and Taylor, E. W. (1969). *Biochemistry*, **8**, 811.

Frieden, C. (1971). *Ann. Rev. Biochem.*, **40**, 653.

Frieden, C. and Alberty, R. A. (1955). *J. biol. Chem.*, **212**, 859.

Frieden, C. and Colman, R. (1967). *J. biol. Chem.*, **242**, 1705.

Frost, A. A. and Pearson, R. G. (1961). *Kinetics and Mechanism*, 2nd Edn, Wiley, New York.

Gerhart, J. C. and Pardee, A. B. (1962). *J. biol. Chem.*, **237**, 891.

Gerhart, J. C. and Schachman, H. K. (1968). *Biochemistry*, **7**, 738.

Gibbons, B. H. and Edsall, J. T. (1963). *J. biol. Chem.*, **238**, 3502.

Gibbons, B. H. and Edsall, J. T. (1964). *J. biol. Chem.*, **239**, 2539.

Gibson, Q. H. (1969). *Methods in Enzymology*, **16**, 187.

Gibson, Q. H. and Parkhurst, L. J. (1968). *J. biol. Chem.*, **243**, 5521.

Guggenheim, E. A. (1926). *Phil. Mag.*, **2**, 538.

Gutfreund, H. (1950). *Progress in Biophys.*, **1**, 1.

Gutfreund, H. (1959). *Disc. Faraday Soc.*, **27**, 229.

Gutfreund, H. (1965). *An Introduction to the Study of Enzymes*, Blackwell, Oxford.

Gutfreund, H. (1969). *Methods in Enzymology*, **16**, 229.

Gutfreund, H. (1971). *Ann. Rev. Biochem.*, **40**, 315.

Gutfreund, H. and Hammond, B. R. (1963). *Nature*, **198**, 667.

Gutfreund, H. and Knowles, J. R. (1967). *Essays in Biochemistry*, **3**, 25.

Gutfreund, H. and McMurray, C. H. (1970). *Biochem. Soc. Symp.*, **31**, 39.

Gutfreund, H. and Sturtevant, J. M. (1959). *Biochem. J.*, **73**, 1.

Hague, D. N. (1971). *Fast Reactions*, Wiley, London.

Haldane, J. B. S. (1930). *Enzymes*, Longmans, Green, London (paperback reprint M.I.T. Press, 1965).

Halford, S. E. (1970). *Ph.D. Thesis*, University of Bristol (also *Biochem. J.*, in press).

Hall, R. L., Vennesland, B., and Kézdy, F. J. (1969). *J. biol. Chem.*, **244**, 3991.

Hammes, G. G. and Hurst, J. K. (1969). *Biochemistry*, **8**, 1083.

Hammes, G. G. and Schimmel, P. R. (1970). *The Enzymes*, 3rd Edn, **2**, 67.

Hammett, L. P. (1940). *Physical Organic Chemistry*, McGraw-Hill, New York.

Hammett, L. P. (1952). *Introduction to the Study of Physical Chemistry*, McGraw-Hill, New York.

Hammond, B. R. and Gutfreund, H. (1955). *Biochem. J.*, **61**, 187.

Heck, H. d'A. (1969). *J. biol. Chem.*, **244**, 4375.

Henderson, L. J. (1913). *The Fitness of the Environment*, Beacon Press, Boston (paperback reprint, 1958).

Hess, B., Haekel, R., and Brand, K. (1966). *Biochem. Biophys. Res. Comm.*, **24**, 824.

Hess, B. and Wurster, B. (1970). *FEBS Letters*, **2**, 73.

Hill, T. L. (1960). *An Introduction to Statistical Mechanics*, Addison-Wesley, Reading, Mass.

Hill, T. L. (1966). *Matter and Equilibrium*, Benjamin, New York.

Hill, T. L. (1968). *Thermodynamics for Chemists and Biologists*, Addison-Wesley, Reading, Mass.

Hinshelwood, C. N. (1951). *The Structure of Physical Chemistry*, Oxford University Press, Oxford, p. 399.

Ho, C. and Sturtevant, J. M. (1963). *J. biol. Chem.*, **238**, 3499.

Janin, J. and Cohen, G. N. (1969). *Europ. J. Biochem.*, **11**, 520.

Janin, J. and Iwatsubo, M. (1969). *Eur. J. Biochem.*, **11**, 530.

Jencks, W. P. (1969). *Catalysis in Chemistry and Enzymology*, McGraw-Hill, New York.

Kabak, H. R. (1970). *Ann. Rev. Biochem.*, **39**, 561.

Katchalski, E. (1970). *Structure–Function Relationships of Proteolytic Enzymes*, Academic Press, New York, p. 198.

Katchalsky, A. and Curran, P. F. (1965). *Non-equilibrium Thermodynamics in Biophysics*, Harvard, Cambridge, Mass.

Katchalsky, A. and Spangler, R. (1968). *Quarterly Rev. Biophys.*, **1**, 127.

Katz, B. (1966). *Nerve, Muscle and Synapses*, McGraw-Hill, New York.

Kauzmann, W. (1959). *Adv. Protein Chemistry*, **14**, 1.

Kauzmann, W. (1966). *Kinetic Theory of Gases*, Benjamin, New York.

Kautzmann, W. (1967). *Thermodynamics and Statistics*, Benjamin, New York.

Kellett, G. L. (1971). *J. Mol. Biol.*, **59**, 401.

Kellett, G. L. and Gutfreund, H. (1970). *Nature*, **227**, 921.

Kilmartin, J. V. and Rossi-Bernardi, L. (1969). *Nature*, **222**, 1243.

Kilmartin, J. V. and Rossi-Bernardi, L. (1971). *Biochem. J.*, **124**, 1243.

King, E. L. and Altman, C. (1956). *J. Phys. Chem.*, **60**, 1375.

Kirschner, K., Eigen, M., Bittman, R., and Voight, B. (1966). *Proc. Nat. Acad. Sci. (U.S.A.)*, **56**, 1661.

Kirschner, M. W. and Schachman, H. K. (1971). *Biochemistry*, **10**, 1900 and 1919.

Kirtley, M. E. and Koshland, D. E. (1967). *J. biol. Chem.*, **242**, 4192.

Klotz, I. M. (1964). *Introduction to Chemical Thermodynamics*, Benjamin, New York.

Klotz, I. M. (1970). *Arch. Biochem. Biophys.*, **138**, 704.

Koshland, D. E. (1960). *Adv. Enzymol.*, **22**, 45.

Koshland, D. E., Nemethy, G., and Filmer, D. (1966). *Biochemistry*, **5**, 365.

Krebs, H. A. (1953). *Biochem. J.*, **54**, 78.

Krebs, H. A. and Roughton, F. J. W. (1948). *Biochem. J.*, **43**, 550.

Levy, H. M., Sharon, N., and Koshland, D. E. (1959). *Proc. Nat. Acad. Sci. (U.S.A.)*, **45**, 785.

Lipmann, F. (1941). *Adv. Enzymol.*, **1**, 99.

Massey, V., Gibson, Q. H., and Veeger, C. (1960). *Biochem. J.*, **77**, 341.

Medawar, P. B. (1969). *The Art of the Soluble*, Penguin, Harmondsworth, p. 45.

Michaelis, L. and Menten, M. L. (1913). *Biochem. Z.*, **49**, 333.

Mildvan, A. L. (1970). *The Enzymes*, 3rd Edn, **2**, 446.

Mitchell, P. D. (1966). *Biol. Revs.*, **41**, 445.

Monod, J., Changeux, J.-P., and Jacob, F. (1963). *J. Mol. Biol.*, **6**, 306.

Monod, J., Wyman, J., and Changeux, J.-P. (1965). *J. Mol. Biol.*, **12**, 88.

Moroney, M. J. (1951). *Facts from Figures*, Penguin, Harmondsworth.

Noble, R. W., Reichlin, M., and Gibson, Q. H. (1969). *J. biol. Chem.*, **241**, 2403.

Palmer, G. and Massey, V. (1968). *Biological Oxidation*, p. 263, Wiley, New York.

Pauling, L. (1970). *Chemistry in Britain*, p. 468.

Perham, R. N. and Anderson, P. J. (1970). *Biochem. Soc. Symp.*, **31**, 49.
Perutz, M. F. (1970). *Nature*, **228**, 726.
Porter, G. (1963). *Rates and Mechanisms of Reactions*, Wiley, New York, p. 1055.
Rabin, B. R. (1966). *Biochem. J.*, **102**, 22c.
Reed, L. J. and Cox, D. J. (1970). *The Enzymes*, 3rd Edn, **1**, 213.
Reynolds, S. J., Yates, D. W., and Pogson, C. I. (1971). *Biochem. J.*, **122**, 285.
Rose, I. A. (1966). *Ann. Rev. Biochem.*, **35**, 23.
Rose, I. A. (1970). *The Enzymes*, 3rd Edn, **1**, 281.
Roughton, F. J. W. (1963). *Rates and Mechanisms of Reactions*, Wiley, New York, p. 703.
Roughton, F. J. W. (1970). *Biochem. J.*, **117**, 801.
Roughton, F. J. W., Otis, A. L., and Lyster, R. L. J. (1955). *Proc. Roy. Soc. (Lond.)*, **B144**, 29.
Scatchard, G. (1949). *Ann. N.Y. Acad. Sci.*, **51**, 460.
Schachman, H. K. (1957). *Methods in Enzymology*, **4**, 32.
Scheraga, H. A. (1963). *The Proteins*, Academic Press, New York, Vol. 1, Chap. 6.
Schwarz, G. (1970). *Eur. J. Biochem.*, **12**, 442.
Segal, H. L. (1959). *The Enzymes*, 2nd Edn, **1**, 1.
Seltzer, S., Hamilton, G. A., and Westheimer, F. H. (1959). *J. Am. Chem. Soc.*, **81**, 4018.
Shore, J. D. and Gutfreund, H. (1970). *Biochemistry*, **9**, 4655.
Silverstein, E. and Boyer, P. D. (1964). *J. biol. Chem.*, **239**, 3901 and 3908.
Stauffer, H., Srinivasan, S., and Lauffer, M. A. (1970). *Biochemistry*, **9**, 193.
Stinson, R. A. and Gutfreund, H. (1971). *Biochem. J.*, **121**, 235.
Suelter, C. H. (1970). *Science*, **168**, 789.
Tanford, C. (1961). *Physical Chemistry of Macromolecules*, Wiley, New York.
Trentham, D. R. and Gutfreund, H. (1968). *Biochem. J.*, **106**, 455.
Trentham, D. R., McMurray, C. H., and Pogson, C. I. (1969). *Biochem. J.*, **114**, 19.
Veech, R. L., Raijman, L., and Krebs, H. A. (1970). *Biochem. J.*, **117**, 499.
Villet, R. H. and Dalziel, K. (1969). *Biochem. J.*, **115**, 633.
Weber, G. and Anderson, S. R. (1965). *Biochemistry*, **4**, 1942.
Weber, K. (1968). *Nature*, **218**, 1116.
Westheimer, F. H. (1961). *Chem. Revs.*, **61**, 265.
Westheimer, F. H. (1963). *Proc. Chem. Soc.*, p. 253.
Weisz, P. B. (1962). *Nature*, **195**, 772.
Wiley, D. C. and Lipscomb, W. N. (1968). *Nature*, **218**, 1119.
Wilson, E. B. (1952). *An Introduction to Scientific Research*, McGraw-Hill, New York.
Winlund, C. C. and Chamberlin, M. J. (1970). *Biochem. Biophys. Res. Comm.*, **40**, 43.
Wurster, B. and Hess, B. (1970). *Hoppe-Seyler's Z.*, **351**, 869.
Wyman, J. (1964). *Adv. Protein Chem.*, **19**, 223.
Wyman, J. (1968). *Quarterly Rev. Biophys.*, **1**, 35.
Wyman, J. and Ingals, E. N. (1943). *J. biol. Chem.*, **147**, 297.

Author Index

233

Subject Index